The Security of Sea Lanes of Communication in the Indian Ocean Region

First published in 2007, this book focuses on the security of sea lanes of communication. It was a joint publication between the Maritime Institute of Malaysia (MIMA) and the Indian Ocean Research Group (IORG) and is an important book for three particular reasons. First, it takes a step forward in identifying key policy themes that can be applied to interstate cooperation around the Indian Ocean Region (IOR). Second, the particular theme discussed is not only central to the economic well-being of Indian Ocean countries, but also to many of the world's most important trading states, and finally the various discussions within the book raise a host of issues to which regional as well as non-regional policy-makers should give serious consideration.

Dennis Rumley is Professor of Indian Ocean Studies and Distinguished Research Fellow at Curtin University, Australia. He was formerly Australia's focal point to the Indian Ocean Rim Academic Group (IORAG) and Vice-Chair from November 2011 to November 2013.

An editorial board member of various international journals, he has also published 11 books and more than 130 papers on political geography and international relations, electoral geography, local government, federalism, Australia's regional relations, geopolitics, India-Australia relations and the Indian Ocean Region. He is currently Chairperson of the Indian Ocean Research Group (IORG) and is Foundation Chief Editor of its flagship journal, *Journal of the Indian Ocean Region*.

Rumley edited a major study on Indian Ocean Security which was launched in Canberra by the Australian Deputy Foreign Minister in March 2013 - *The Indian Ocean Region: Security, Stability and Sustainability in the 21ˢᵗ Century*. He has also recently edited *The Political Economy of Indian Ocean Maritime Africa*, co-edited *Indian Ocean Regionalism*, with Timothy Doyle, and republished his co-edited book with Julian Minghi, *The Geography of Border Landscapes*, in a new *Routledge Library Editions: Political Geography series*.

Dr. Sanjay Chaturvedi is Professor of Political Science at the Centre for the Study of Geopolitics, Panjab University, India. He was awarded the Nehru Centenary British Fellowship, followed by the Leverhulme Trust Research Grant, to pursue his post-doctoral research at University of Cambridge, England. He was also a Third Cohort Fellow of the India-China Institute at the New School, New York (2010-2012) and was the Co-Chair of the Research Committee on Political and Cultural Geography (RC 15) of International Political Science Association (IPSA) from 2006 to 2012.

Founding Vice-Chairman of the Indian Ocean Research Group (IORG Inc.), Chaturvedi is the Co-Editor of the *Journal of the Indian Ocean Region* and the Regional Editor of *The Polar Journal*. He also serves on the Advisory Board of *Geopolitics* and *Journal of Borderlands Studies*.

Chaturvedi has served on the Indian delegations to the Antarctic Treaty Consultative Meetings (ATCMs) since 2007 and Track II Trilateral Dialogue on the Indian Ocean (TDIO) among India, Australia and Indonesia. His recent co-authored books include *Climate Terror: A Critical Geopolitics of Climate Change*, with Timothy Doyle, *Climate Change and the Bay of Bengal: Emerging Geographies of Hope and Fear*, with Vijay Sakhuja.

Mat Taib is a retired Captain turned security analyst after 28 years of service in the Royal Malaysian Navy. He has penned dozens of documents, research reports and articles and keynote addresses for VIPs including the Prime Minister. He also commanded and assumed the coveted position of Director of Naval Strategic Plan.

He is a Navigation Specialist. Upon being awarded a distinguished graduate of the Armed Forces Staff College, he won a scholarship to study at the National University and later Lancaster University, UK, earning him an MA in Science and Technology.

His second career at the Maritime Institute of Malaysia involved substantive research on the security of Sea Lanes, particularly the Straits of Malacca, Indian Ocean and South China Sea. Some of these works have been published.

Taib's expertise was recognized through assignments such as being the technical adviser to the National Security Council and Head of Malaysian-Indonesia (MALINDO-INCSEA) delegation.

He is currently a part-time independent consultant.

ARC ACKNOWLEDGEMENT

The editors would like to acknowledge the funding of the Australian Research Council Discovery Project DP120101166: "Building an Indian Ocean Region" for their generous assistance in the conduct and completion of this research, writing and publication 2012-2015.

The Security of Sea Lanes of Communication in the Indian Ocean Region

Edited by

Dennis Rumley, Sanjay Chaturvedi and Mat Taib Yasin

Routledge
Taylor & Francis Group

First published in 2007

This edition first published in 2015 by Routledge
2 Park Square, Milton Park, Abingdon, Oxon, OX14 4RN
and by Routledge
711 Third Avenue, New York, NY 10017
Routledge is an imprint of the Taylor & Francis Group, an informa business

Publisher's Note
The publisher has gone to great lengths to ensure the quality of this reprint but points out that some imperfections in the original copies may be apparent.

Please note, the original 'List of Contributors' is incomplete and several contributors are absent. Those absent are Noor Apandi Osni, Dennis Rumley, Vijay Sakhuja, Swaran Singh, Mat Taib Yasin, Mokhzani Zubir.

Disclaimer
The publisher has made every effort to trace copyright holders and welcomes correspondence from those they have been unable to contact.

A Library of Congress record exists under LC control number: 2010337283

ISBN 13: 978-1-138-92912-8 (hbk)
ISBN 13: 978-1-315-68136-8 (ebk)
ISBN 13: 978-1-138-92917-3 (pbk)

The Security of Sea Lanes of Communication in

the

Indian Ocean Region

edited by

Dennis Rumley, Sanjay Chaturvedi and Mat Taib Yasin

MARITIME INSTITUTE OF MALAYSIA
INSTITUT MARITIM MALAYSIA

Published by
Maritime Institute of Malaysia (MIMA)
Unit B-6-8 Megan Avenue II
12 Jalan Yap Kwan Seng
50450 Kuala Lumpur
Malaysia
Tel: +603 216 12 960
Fax: +603-216 14 035
E-mail: mima@mima.gov.my
Website: http://www.mima.gov.my/

ISBN 978-9275-40-7

The views expressed in this book are entirely that of the authors and do not reflect the views of MIMA.

CONTENTS

FOREWORD
PREFACE
CONTRIBUTORS

THE GEOSTRATEGIC IMPORTANCE OF THE INDIAN OCEAN

THE INDIAN OCEAN AS A GLOBAL TRADING AND ENERGY LIFELINE

SECURITY THREATS IN THE INDIAN OCEAN

CHALLENGES AND PROSPECTS FOR MARITIME SECURITY COOPERATION

FOREWORD

It gives me very great pleasure to write this Foreword for the joint publication between the Maritime Institute of Malaysia (MIMA) and the Indian Ocean Research Group (IORG) based on a Conference hosted by MIMA and held in Kuala Lumpur in July 2005. This is an important book for three particular reasons. First, it takes another step forward in identifying key policy themes that can be applied to interstate cooperation around the Indian Ocean Region (IOR). Second, the particular theme discussed in the volume - the security of sea lanes of communication - is not only central to the economic well-being of Indian Ocean countries, but also to many of the world's most important trading states. Third, the various discussions within the book raise a host of issues to which regional as well as non-regional policy-makers should give serious consideration. From this perspective, I sincerely hope that this volume enjoys a wide Indian Ocean regional as well global circulation.

Like all such collaborative efforts, this volume would not have been possible but for the contribution of so many individuals at various stages of its production. As Conference host, MIMA is especially grateful to the Honourable Minister of Transport, Malaysia, Dato' Seri Chan Kong Choy, who made time to officiate the Conference; Chairman of MIMA, Dato' Seri Ahmad Ramli Mohd Nor; and also MIMA staff who spared no effort to ensure that the conference ran smoothly, We are also thankful to the Royal Malaysian Navy, which organised an excellent field excursion in the Straits of Malacca. We are greatly indebted to the IORG, and especially its coordinators, Sanjay Chaturvedi and Dennis Rumley for bringing the Conference to Malaysia and to all of the international and local speakers and participants who helped make the Conference such an outstanding success.

Last, but not least, on behalf of MIMA, I would like to express my sincere gratitude to all of the contributors to this volume for their expertise, patience and ability to analyse and discuss so many important policy issues related to the security of sea lanes of communication in the IOR. Furthermore, I would like to congratulate Iskandar Sazlan Mohd Salleh, Senior Researcher and Prof. Dr Mohd Ibrahim B Hj. Mohamad, Director of Research, MIMA, for co-ordinating the publication of this volume. I am certain that the results will make a significant contribution to the policy debate in the Indian Ocean Region.

Cheah Kong Wai,
Director-General

PREFACE

This is the third volume produced by the Indian Ocean Research Group (IORG - http://www.iorgroup.org) and is based on a selection of papers given at a Conference very generously hosted by the Maritime Institute of Malaysia (MIMA) in Kuala Lumpur in July 2005 and is also published by MIMA. The Conference organisers are grateful to many people who contributed so generously to its success. In particular, thanks are due to all of the first-rate staff of MIMA who ran the Conference so smoothly and efficiently and with such good humour. We all owe a special debt to the Chairman of MIMA, Dato' Seri Ahmad Ramli Mohd Nor, to the Director-General, Dato' Cheah Kong Wai, and to local organiser, Captain Mat Taib Yasin.

The Royal Malaysian Navy organised a wonderfully unforgettable full-day field trip to and through the Straits of Malacca and we are all especially grateful to them. The Indian High Commissioner to Malaysia, His Excellency R. L. Narayan, hosted a sumptuous reception, as did the Australian High Commissioner, His Excellency Mr James Wise. We are also grateful to Mr Simon Merrifield, Deputy High Commissioner of Australia, for being an excellent stand-in host and for assisting in the launching of the IORG second volume, *Energy Security and the Indian Ocean Region*.

As always, the Conference could not have taken place without the participation of our excellent and growing network of international scholars. We thank all of them for their contributions, their continued support, their enthusiasm, their genuine commitment to Indian Oceanness and for their patience. We hope that this volume, like all IORG volumes, can make a useful contribution to the policy debate on key issues of concern to all inhabitants of the Indian Ocean Region.

Dennis Rumley, Sanjay Chaturvedi and Mat Taib Yasin

CONTRIBUTORS

Mohd Nizam Basiron, Centre for Coastal and Marine Environment, Maritime Institute of Malaysia (MIMA), Kuala Lumpur

Sam Bateman, Senior Fellow, Maritime Security Programme, Institute of Defence and Strategic Studies (IDSS), Singapore and Professorial Research Fellow, Centre for Maritime Policy, University of Wollongong, Australia

Robert Beckman, Associate Professor, Faculty of Law, National University of Singapore

Aparajita Biswas, Director for the Centre for African Studies, University of Mumbai, India

Christian Bouchard, Associate Professor, Laurentian University, Sudbury, Canada

Sanjay Chaturvedi, Reader in Centre for the Study of Geopolitics, Panjab University, Chandigarh, India

Timothy Doyle, Professor of Political Science, University of Keele, UK

Hasjim Djalal, Ambassador-at-Large, Djakarta, Indonesia

Vivian Louis Forbes, Map Librarian, University of Western Australia and Adjunct Associate Professor, Curtin University, Perth

Manoj Gupta, Australian Defence Force Academy, University of New South Wales, Canberra

Joshua Ho, Senior Fellow, Institute of Defence and Strategic Studies, Singapore

Jean Houbert, Honorary Research Fellow, University of Aberdeen, Scotland

Siti Norniza Zainul Idris, Researcher, Centre for Ocean Law and Policy, Maritime Institute of Malaysia (MIMA), Kuala Lumpur

Nazery Khalid , Maritime Institute of Malaysia (MIMA), Kuala Lumpur

Prologue

KEYNOTE ADDRESS BY THE HONOURABLE CHAN KONG CHOY, MINISTER OF TRANSPORT MALAYSIA
(The Legend Hotel, Kuala Lumpur, 11th July 2005)

It is indeed a great pleasure for me to present this keynote address at this international conference on "Sea-lane Security in the Indian Ocean". The hosting of this conference, anchored on the broader premise of the term "security", is very timely and would allow discussions to proceed beyond the realm of traditional security concerns. With the presence of both foreign and local experts from numerous disciplines, I am confident that this conference will be a stimulating and productive meeting of the minds on one of the world's major strategic waterways. I bid all of you a warm Selamat Datang, particularly to our foreign speakers and participants. Welcome to Malaysia.

Economic and security needs are amongst the basic human needs listed in Maslow's hierarchy of needs. Indeed, these needs form the essence of comprehensive security and are intertwined in such a way that economic prosperity is only possible in an environment of peace and security. Conversely, enduring peace and security can only be ensured if there is economic prosperity. For this reason, most, if not all, human endeavours and political agendas are directed towards the attainment of comprehensive security.

Throughout history, the sea has been an important conduit for economic prosperity. The waterways have been used as the primary medium of trade and have provided the impetus for the growth of maritime enterprises. The growth in the dependency on the sea as an economical and efficient means of transportation is well acknowledged. The United Nations Conference on Trade and Development (UNCTAD) has noted in its report entitled, "Review of Maritime Transport 2004", that world sea-based trade recorded a steady annual increase for the last 20 years. In terms of global percentage of breakdown of seaborne loaded goods by continent in 2003, Africa accounts for 8.9 percent, America 20.7 percent, Asia 37.2 percent, Europe 25.1 percent and others 8.1 percent. Given these developments, sea-based trade is poised to grow in size and importance, hence underlining the need to focus on the safety and security of sea-lanes of communications (SLOCs).

The Indian Ocean is the third largest ocean in the world. It covers an area of about 78 million square kilometers, comprising some 20 percent of the world's total sea area. It includes, amongst others, the Red Sea, the Persian Gulf, the Arabian Sea, the Bay of Bengal and the Andaman Sea. The Indian Ocean is surrounded by land on three sides. On the west by the African Continent, the north by Southwest Asia and the south by Western Australia. The number of sovereign states in the Indian Ocean Region can be put at 47, including 30 littoral states, 5 islands and 12 land-locked countries. In addition, France and the United Kingdom are amongst the non Indian Ocean states that possess island territories in the region.

Today, the Indian Ocean represents at least three different trade systems. First, the extra-regional bustling trade transit between East and West, or between the Pacific and the Atlantic, passing through the Indian Ocean. The second trade system facilitates natural oil and gas shipments coming from the Persian Gulf. The third trade system is regional or sub-regional trade that serves to distribute goods either in the Bay of Bengal or the Arabian Sea, from outside or from within the Indian Ocean and feeding local goods into the larger trading system.

As a result, the Indian Ocean is one of the busiest waterways in the world. The intensity of trading activities through the Ocean is clearly evident from the volume of ships passing through and the growth of regional ports. With over 100,000 ships carrying various cargoes passing through annually, the Indian Ocean is truly the artery of the global economy. While serving and facilitating Asian economic giants like Japan, South Korea, China and India, it plays a crucial role in the economic well-being of the United States and Europe. In terms of energy security, almost all the oil transportation sea-lanes from West Asia to the rest of the world pass through the Ocean. This includes the daily transshipment of about 11 million barrels of oil from the Middle East to East Asia, the world's largest oil-consuming region.

Basing on the current process of industrialization in emerging giant Asian economies like China and India, the demand for energy is going to increase by leaps and bounds. For India alone, the demand for crude oil is projected to increase from the current 90 million tons to 720 million tons in 2020. The demand for LNG during the period is projected to increase from 2.5 million tons to 200 million tons.

The Indian Ocean is now traversed by all of the world's major shipping lines. Vessels owned by Maersk, P&O, Evergreen, COSCO, MSC, NYK and other major lines link major eastern ports of Kaoshiung, Shanghai, Busan, Singapore, Port Klang, Luang Prabang to principal European ports of Rotterdam, Hamburg and Antwerp through the Indian Ocean.

A number of ports within the Indian Ocean rim are now emerging to be star players in international commerce. While on the western rim of the Indian Ocean lies the ports of Dubai, Karachi, Muscat, Mumbai and Jeddah, on the eastern rim are the ports of Chittagong, Singapore and Malaysia's own Port Klang and Port Tanjong

Pelepas, both of which have emerged to be amongst Asia's top ten container ports.

The increase in regional container trafficking movement marked an increase in intra-regional trade. In 2004/2005, the volume of India's container traffic totaled 4.15 million TEUs. They are largely shipped through major regional ports like Singapore, Colombo and Dubai.

The freedom of navigation and free flow of trade facilitated by the Ocean have directly contributed trillions of dollars to the Gross Domestic Products (GDP) and millions of jobs to the respective giant economies. Indirectly, millions more jobs are created in countries that depend on the economic strength of these powerful states. Any interference to the freedom of navigation through the Ocean that results in the interruption of the flow of trade and oil supplies will thus bring about a catastrophic impact on the global economy.

Globalization and the growing dependence of states on foreign trade have certainly raised the profile and increased the importance of the Indian Ocean SLOCs. Maritime transport, as the most economical and effective means of transport to support international trade, will correspondingly see an increase in activities in the Indian Ocean. Coincidently, the increase in the number of users of the sea-lanes has also created more opportunities for those with ill intentions to carry out their predatory acts, such as piracy and ship-jacking. These acts, committed mostly in narrow congested waterways adjoining the Indian Ocean have drawn major international concern. To countries like Malaysia, of equal concern is the threat posed by trans-boundary criminal activities operating from the Andaman Sea, on the Northeast of the Indian Ocean. Reports have indicated that this area is being used as a hub for the trafficking of firearms, drugs and illegal immigrants.

As economic prosperity is the foundation to regional security and stability, regional states must endeavour to enhance the level of their existing intra-regional economic cooperation. At the same time, new cooperative avenues must be explored. Increasing regional cooperation such as those initiated on the platform of the Indian Ocean Region-Association for Regional Cooperation (IOR-ARC) will enable Indian Ocean countries to participate more actively and effectively in the global economy and leverage their membership in the grouping.

While regional economic and security cooperation is the ideal and the only way forward, in the pursuance of this aspiration, we may have to endure numerous obstacles and challenges given our regional diversities. The creation of a suitable regional political climate is one of the fundamental needs. Towards this end, existing fear and suspicion between and amongst all regional states and international stakeholders must be removed and the creation of new ones must be avoided. Hence, the politics of "divide and engage" that is perceived to isolate or deny the Indian Ocean to any particular stakeholder must be avoided. It surely will make genuine cooperation impossible.

In managing restricted waterways like the Strait of Malacca, close cooperation between littoral states and the international users is a matter of necessity, not a choice. Although it is to the economic interest of all stakeholders that the waterway remains open and secure for the movement of commerce, to the littoral states the waterway means more than that. Apart from the flow of commerce and trade, the waterway is a source of economic income, livelihood, food, as well as national sovereignty and security. Thus, for them, the definition of security is all-encompassing Ð physical or territorial security, economic security, food security and environmental security. Any cooperative formula proposed must therefore take cognizance of this.

Given such a reality, it is imperative for this august gathering to find solutions to ensure that the Indian Ocean continues to be a safe and secure waterway to be continuously shared by shippers, fishermen and holiday-makers. In this regard, I am glad that dialogues are ongoing to explore how littoral states of the Indian Ocean can join hands to enhance their economic and security cooperation that will ultimately ensure that regional waterways remain safe for the movement of international trade. As I understand it, one of the main objectives behind the formation of the Indian Ocean Research Group (IORG) is to initiate a policy-oriented dialogue, in the true spirit of partnership, among governments, industries, NGOs and communities, towards realizing a shared, peaceful, stable and prosperous future for the Indian Ocean Region. To build on this further, the initiatives for long-term cooperation among the principal players should be based on mutual trust. All stakeholders must endeavour to understand and respect each other's different levels of dependencies and concerns over the Indian Ocean. Economic and security cooperative initiatives must therefore be pursued hand-in-glove as both are complementary to one another. These initiatives must be preceded by a harmonious political environment.

ON THAT NOTE, LADIES AND GENTLEMEN, I NOW OFFICIALLY DECLARE OPEN THE THIRD IORG CONFERENCE.
Thank you.

Chapter 1

Securing Sea Lanes of Communication in the Indian Ocean Region

Dennis Rumley, Sanjay Chaturvedi and Mat Taib Yasin

Introduction

The main purpose of this book is to present a critical discussion of the state of current regional policy and practice on the security of sea lanes of communication in the Indian Ocean. Its primary aim is to invigorate the debate, to inform policy-makers and to provide a basis upon which individual state and collective policies might change in order to minimise insecurity in the Indian Ocean Region (IOR). The book is divided into four principal sections each containing four chapters. The first section deals with the geostrategic importance of the Indian Ocean. The second section examines the Indian Ocean as a global trading and energy lifeline. The third section evaluates some of the main maritime security threats and the final section considers the challenges and prospects for greater regional maritime security cooperation.

The principal purpose of the present chapter is to provide a succinct overview of the contribution of each of the subsequent chapters to an understanding of the complexity and importance of sea lanes of communication in the Indian Ocean Region. However, before doing so, some basic security considerations relevant to any geopolitical analysis of sea lanes of communication will be discussed.

Sea Lanes of Communication: Some Fundamental Security Considerations

It goes almost without saying that the security of sea lanes (also referred to as sea *lines*) of communication (SLOCs) is vital to the functioning of the global economy. In 2001, 80% of world trade by value and 90% by volume was in the form of sea trade and this involved 1.2 million seafarers, 46,000 vessels and 4,000 ports (Harrald, 2005, 160). Maximising the economic security of sea trade thus necessitates the maximisation of security within and among all five basic elements in the maritime trading process - seafarers (for example, background and nationality), vessels (for example, registration and seaworthiness), cargoes (for example, nature and destination), ports and SLOCs. However, any understanding of the security of SLOCs cannot be fully appreciated in isolation from the overall maritime security environment.

Interestingly, there appears to be no commonly accepted definition of what constitutes maritime security that might be used as a basis for regional cooperation

(Banlaoi, 2005, 59). To date, the collective maritime security environment has been conceptualised as a composite of sea power and the naval arms build up, island and maritime boundary issues, navigational regimes, activities in the EEZ, competition over resources and the maintenance of law and order at sea, including the protection of SLOCs (Paik, 2005). To these essential elements can also be added others associated primarily with the stability and integrity of ocean littoral states, the insecurity role of non-state actors and the question of maritime environmental insecurity. Furthermore, the achievement of great power status has long been associated with control over the seas and especially with the maintenance of an internally secure set of sea lanes of communication within imperial domains at various scales (for example, Mahan, 1890). Such sea power was not only essential to obtain and sell commodities which reinforced state economic power, but also exemplified various designs and strategies over the use and control of maritime space. In addition, sea power was necessary for the perpetuation of a system of flows and linkages that systematically disadvantaged peripheral states and reaffirmed maritime hegemony. Non-negotiated 'transgressors' into this taken-for-granted maritime world were perceived to be and represented as threats to national and international security. Changes to the structure of sea power were thus to be avoided and 'collectively' confronted by states only when hegemonic interests were threatened through a process of 'collective power coattailing'.

Some commentators differentiate the term 'sea power' from 'naval mastery', with the latter implying influence at a global rather than a regional level and indicating:

"...a situation in which a country has so developed its maritime strength that it is superior to any rival power, and that its predominance is or could be exerted far outside its home waters, with the result that it is extremely difficult for other, lesser states to undertake maritime operations or trade without at least its tacit consent" (Kennedy, 1983, 9).

The condition of naval mastery throughout recent history is applicable to a very few states, and was possessed by Britain in the nineteenth century and clearly is now possessed by the United States. As has been pointed out, the United States relies on the control of the oceans and SLOCs to implement its strategic objectives, but in order to facilitate these objectives in the IOR necessitates access to appropriate maritime infrastructure such as basing facilities, ports, and, of course, vessels (Stasinopoulos, 2003). Realising IOR strategic objectives thus requires the United States to engage in bilateral cooperation with regional states and regional navies. Furthermore, from a US as well as a regional perspective, maritime security has assumed a much higher level of priority post-9/11.

Up until relatively recently, much of the sea power debate has concentrated almost exclusively on its military definition and thus states have been concerned with the development of maritime strategies and maritime security policies which ignore or underplay a wide range of non-military considerations. From a military perspective, for example, "a modern maritime strategy involves air, sea and land forces operating jointly to influence events in the littoral together with traditional blue water maritime concepts of sea denial and sea control" (Australian Parliament, 2004, 8). However, in addition to the military component, a much more broadly-based maritime security strategy would necessarily incorporate a wide range of economic, environmental, political and social considerations and thus require greater inter-organisational collaboration within states and the amelioration of "bureaucratic sclerosis" for its successful implementation (Roy-Chaudhury, 2000, 187; Australian Parliament, 2004, 12). Furthermore, while ocean littoral states will endeavour to develop their individual maritime security strategies, in the final analysis, securing the maritime environment, which among other things involves the building of an internationally stable maritime regime as well as the implementation of maritime confidence-building measures, at a minimum will require regional and even global cooperation in the twenty-first century (Grey, 1993; Singh, 1993). However, balancing maritime security with freedom on the high seas will necessitate a complex and delicate process of international negotiation (Young, 2005).

The Geostrategic Importance of the Indian Ocean

It was suggested in an earlier IORG volume that there are at least five interrelated reasons why Western observers have generally tended to underestimate the geostrategic importance of the Indian Ocean Region (Rumley and Chaturvedi, 2004, 23-4). First, since some commentators see the Region as primarily comprising Third World states, then by association the Region has been accorded a lower level of significance compared with other Oceans, especially the Atlantic and Pacific Oceans. Second, the Northern-centric primarily English-based system of global knowledge contributes to a generalised level of ignorance about the Region and about regional states. Northern-Western resources are unlikely to support Indian Ocean educational programmes when better economic returns are perceived to exist elsewhere. Third, since the Region is a former arena of European colonial competition, to some degree, residual core-periphery values of social, economic and political superiority and security dependency continue to prevail. Fourth, there is a relative paucity of reliable writing from students of international affairs on the geostrategic importance of the Indian Ocean Region. Finally, the geopolitical orientation of many Indian Ocean regional states has tended to be away from the Ocean and has been directed internally or towards states and regions of perceived greater geoeconomic significance. However, in the post-Cold War period, and especially in the 21st century, there has been a growing realisation of the increasing geostrategic importance of the Indian Ocean Region. The end of the Cold War, increasing globalisation, India's opening up and Look East policy, the end of apartheid in South Africa, Australia's Look West policy and the increasing concern over an array of non-traditional security threats place new emphasis on the Indian Ocean environment in its broadest sense and, especially, the secure and sustainable use and management of that environment both on and in, as well as below, the Ocean itself.

In terms of international trade, the Indian Ocean is now the world's most important routeway. What this means, among other things, is that the Northern economic powers and the developing world economic powers of China and India (and each of their respective navies) possess a legitimate interest in the security of SLOCs, as well as a special concern for the stability and geopolitical orientation of states proximate to the entrances and exits to the Ocean. In this regard, apart from the troubled Horn of Africa, no less than ten regional states are critical to the free flow of global sea trade - Australia, Indonesia, Iran, Malaysia, Mozambique, Oman, Singapore, South Africa, United Arab Emirates and Yemen. Since these states represent more than half of its membership, then this means that IOR-ARC is a potentially extremely important regional grouping for the construction of a cooperative maritime security regime in the Indian Ocean Region. This implies the necessity of broadening the agenda of IOR-ARC, not only to breathe new life into economic cooperation, but, more importantly, into a wide array of potentially mutually-beneficial cooperative endeavours.

In Chapter 2, Jean Houbert discusses the emergence of three European settler-colonies that became settler-states - South Africa, Australia and Israel - and their implication in the Indian Ocean geopolitical strategies of first Britain and then the United States. The hegemonic power of the United States has not only been entrenched in the Indian Ocean through its base at Diego Garcia, but also by a set of alliances with the three settler-states which were located along important sea lanes of communication. As Houbert points out, of these three states, only Australia has succeeded in solving its "native problem". That is, in Australia, the settlers were able to exert sufficient control over land to ensure the marginalisation of the indigenous minority. In South Africa, in contrast, demographic and political realities meant that solving the "native problem" was never viable. The settler-state of Israel, on the other hand, while being more successful than South Africa in its "native problem" solution, still possesses nothing like the legitimacy and security of Australia.

In terms of the security of Indian Ocean SLOCs, this differential "success" hasmeant that, on the one hand, while the "natives" are in control in South Africa,

Cape route acquires a considerable degree of stability. Furthermore, Australian regional cooperation coupled with US sea hegemony will likely sufficiently deter threats to eastern Indian Ocean choke points. On the other hand, the failure to find an acceptable solution to the settlers-natives conflict in Palestine has a detrimental impact on the Suez canal route. Perceived occupation necessarily generates overt resistance.

Hasjim Djalal in Chapter 3 argues that the Indian Ocean has considerable and increasing importance for Indonesia. From a strategic perspective, Indonesia is located between two oceans and two continents; its geography is sea-oriented; it is rich in natural resources; it has close cultural ties with Indian Ocean states; it is a large and moderate Islamic democracy; it is the birthplace of the Bandung Spirit; and, it was a founder member of Non-Alignment. Politically, Indonesia has been active in developing cooperation in the Indian Ocean Region partly because the Ocean is seen as a "bridge" between Africa and Asia. However, Djalal points out that the process of regional construction has lost some of its momentum, despite the creation of regional institutions such as the Indian Ocean Marine Affairs Cooperation (IOMAC), the Indian Ocean Region-Association for Regional Cooperation (IOR-ARC) and Indian Ocean Tourism Organisation (IOTO). From an economic viewpoint, while Indonesia has an important interest in fisheries, it is neither a member of the Indian Ocean Tuna Commission (IOTC) nor a signatory to the Convention on the Conservation of the Southern Blue Tuna (CC-SBT). The author raises several other policy-relevant issues associated with the need for Indonesia to determine its continental shelf and its contiguous zone, as well as the status and use of the EEZ and the high seas.

Djalal sees the Indian Ocean as the "Ocean of the future", especially since the issue of energy security will ensure its growing global geostrategic significance as an energy route. This will also mean that the geopolitical role of Indonesia is in the process of becoming globally more significant.

In Chapter 4, Swaran Singh analyses the emerging centrality of SLOCs in China's 21st century maritime strategy. He notes that China faces something of an intractable problem due to its rapid economic development creating a consequential increasing dependence on the free flow of goods and services, which makes it potentially vulnerable to the intervention of external powers and non-state security threats. As a result, China has an inherently central interest and an economic necessity for ensuring freedom of navigation in Indian Ocean SLOCs. However, this has placed it in direct competition with other powers which has compounded its policy options and poses important dilemmas for Chinese policy-makers. Singh examines China's alternative strategies in response to these dilemmas and discusses the concept of the Kra Canal Project in Thailand as an alternative to the Malacca Straits.

Since the 1980s, China's growing awareness of its maritime rights has resulted in a shift of strategy to one of "offshore defence". Associated with this change has emerged a debate centred on the Malacca Straits Dilemma (MSD) thesis. In essence, this view states that certain major powers are determined to control the Malacca Straits to China's disadvantage. The Chinese policy response to this has been a three-pronged strategy aimed at reducing import dependence, finding alternative routes and building a credible naval force capable of defending Chinese SLOCs. In addition, China has been strengthening economic engagement with Indian Ocean littoral states in relation to trade, infrastructure and the building of naval facilities (the "string of pearls").

Christian Bouchard in Chapter 5 points out that even though the Mozambique Channel is an increasingly important international waterway, it has generally received insufficient attention from Indian Ocean geopolitical analysts. He suggests that perhaps one of the principal reasons for this is that any Channel closure has to date been regarded as improbable and as not likely to result in any serious disruption to ocean shipping. The Mozambique Channel Maritime Region (MCMR) contains one of several Indian Ocean sea lanes linked to the Cape route and it has been dominated by a French naval presence and contains good fishing grounds. Since local states do not possess a fishing fleet, the MCMR has also been of interest to many non-regional states. As Bouchard argues, the region is in the process of significant change likely resulting in a more conflictual local environment. For example, the region is now attracting the interest not only of the South African navy but also those from China, India and the USA. Furthermore, there are many unresolved territorial issues within the region. Perhaps the most important likely regional change in the future, however, relates to the potential for oil and gas. It is possible that greater regional exploration will reveal very significant oil deposits that, in turn, will have major regional environmental and economic impacts. Consequently, the geopolitical significance of the MCMR will fundamentally change. Meanwhile, low levels of human development and high poverty levels provide a favourable environment for increased criminal activity and for regional non-state security threats.

The Indian Ocean as a Global Trading and Energy Lifeline

The increasing use of the Indian Ocean as a routeway takes on a new significance, especially in a global environment of increasing competition over energy and one in which non-state security threats have become increasingly prevalent. In this fundamentally changed geopolitical context, securing the sea lanes of communication has become an issue of global importance and thus requires a collective response. In particular, the ways in which ocean cargo is perceived on the part of regional and extra-regional states can be of critical security importance.

For example, flows of oil will generally be seen as beneficial since they are among the principal drivers of economic growth. Other energy sources - for example, uranium - may yield both positive as well as negative perceptions depending on environmental ideology as well as the source, destination and ultimate use.

Movements of nuclear waste or enriched plutonium or nuclear weapons, on the other hand, would generally be regarded negatively, not only because of the potential to enhance ocean environmental insecurity in the event of deliberate dumping or accidental spillage, but also due to concerns over nuclear proliferation by state and non-state actors. In all circumstances, the maximisation of ocean security in its widest sense requires the strict application of mutually agreed cargo guidelines and for the enforcement of mutually agreed sanctions in the event of contravention. In order to maximise regional security, all such guidelines need to be mutually agreed and need to give due respect to state sovereignty.

In chapter 6, Nazery Khalid describes the role of the Indian Ocean in promoting global maritime trade and provides an assessment of the importance of the Ocean to world commerce. While the author notes that the Ocean already contains some of the most important global sea lanes, it is likely that, with increasing globalisation, it will benefit from increasing trade flows. Greater trade, in turn, will enhance the demand for energy and thus facilitate an increasing flow of commodities through the Indian Ocean. The region's large population (especially in South Asia) and increasing levels of economic growth, especially in the 'triangle' economies of India, South Africa and Australia, will also help fuel an expansion in consumer demand and thus intra-regional trade. The increasing level of Indian Ocean traffic will stimulate an expansion in regional investment in port infrastructure and an associated growth in regional shipbuilding, particularly in South Asia and Australia. Nevertheless, intra-regional trade still remains at a relatively low level. Nazery Khalid argues that, in order for maritime trade to be enhanced, regional states need to decentralise and privatise their port services and improve infrastructure in order to attract further foreign investment. Taken together, all of these factors will continue to facilitate the use of the Indian Ocean both as a regional as well as a global trade routeway and thus, in the future, Indian Ocean sea lanes will undoubtedly become increasingly important to world trade.

Dennis Rumley and Timothy Doyle in chapter 7 argue that the Indian Ocean is in the process of being transformed into a nuclear ocean since it is becoming increasingly important as a routeway for the transportation of uranium ore, enriched uranium and nuclear waste. They are especially concerned over the environmental security implications of this shift as well as the prospect of increased regional smuggling of radioactive materials and illegal dumping of

The increasing demand for nuclear energy includes those states which are energy dependent (for example, India), those which are industrialising (for example, China) and those involved in the development of nuclear technology (for example, Iran). Growing regional and global interest in nuclear energy is also being propelled by a belief in its "greenhouse friendly" status. In 2006, in the Indian Ocean Region, the number of planned or proposed nuclear reactors in five states (India, Indonesia, Iran, Pakistan and South Africa) represented 39 per cent of the global total. Australia is still debating its position on nuclear energy. Nonetheless, the proposed rapid regional and global growth in the production of nuclear energy in turn places pressure on regional uranium supply states, which are "uranium rich" but are currently "underproducing". The question of the storage and movement of nuclear waste is probably the most difficult and controversial issue in the Indian Ocean nuclear trade and the authors point out that Australia is still seen by many as potentially the most suitable global waste repository.

In chapter 8, Vivian Louis Forbes describes the structure of SLOCs by which shipments of uranium and nuclear waste flow through the Indian Ocean. Forbes concludes that there is no record that the 'internal' trans-Indian Ocean route has actually been used for the transport of nuclear waste. Rather, the records show that due in part to pressures from Indian Ocean states, plutonium (1992) and nuclear waste (1997) were shipped via the Cape of Good Hope and around the southern coast of Australia. The author also critically reviews the current body of international law in relation to this and identifies conflicting views over rights of passage. It is the belief of Indian Ocean littoral states, however, that the 1982 LOS Convention provides them with some basis for the regulation of ships carrying nuclear materials. Such regulation is bolstered by the INF code which now controls the safe carriage of dangerous goods by all vessels. Forbes examines the actual practice of eight Indian Ocean states in relation to this body of law and to other regional agreements. He also refers to a recent Indian Supreme Court case in which it was decided that a Danish vessel found to be carrying hazardous material should be 'driven out' of Indian waters. Based on his evaluation, the author's overall conclusion is that the nuclear transportation industry has had a comparatively unblemished track record, since there has never been a dangerous radioactive spill or loss recorded since the 1960s during the transportation of nuclear materials.

Noor Apandi Osnin and Siti Norniza Zainul Idris in chapter 9 highlight the expansion of the natural gas industry throughout the world as well as the increasing importance of the Indian Ocean as one of the routeways for LNG transshipment. In addition, the authors undertake a detailed evaluation of the security concerns involved in the ocean transportation of LNG and the manner in which the industry has responded to these issues. They conclude that the safety and security of LNG transhipment has increased based on a number of recent developments. First, there has been a radical development in the nature of the LNG carriers themselves due to improvements in the quality and maintenance of these vessels. In addition, the complexity of the vessels minimises the adverse effects of any mishap during LNG transshipment from the liquefaction to the regasification plant. The safety record of LNG carriers in the past 40 years has reassured the industry players that LNG provides less risk in terms of environment and human hazard and minimum loss projection in trade. The authors point out that there has indeed been an improvement in overall operational safety in the past 40 years of operation. Effective training for the personnel, in relation to detailed procedures for loading and unloading the cargo to the storage plant, unique regulations by each state and security and prevention measures are all in place. It seems that the industry has taken all necessary steps to maximise the safety and security of the LNG industry.

Security Threats in the Indian Ocean

To a considerable degree, maritime insecurity will be a reflection of the degree of stability of Ocean states and sub-regions. Over the last decade, much of the Indian Ocean Region has been portrayed by geopolitical analysts as exhibiting increasing levels of instability. For example, most of the West Asian portion of the Indian Ocean Region has been characterised as "*the* (emphasis added) global zone of percolating violence" and "is likely to be a major battlefield, both for wars among nation-states and, more likely, for protracted ethnic and religious violence" (Brzezinski, 1997, 52-3). In addition, most of the Northern half of the Indian Ocean Region, including India, has been incorporated into a US-designated "Southern Belt of Strategic Instability" which stretches from southern Japan in the east to northern Italy in the west (Flanagan, Frost and Kugler, 2001, 17). Not only do such constructions tend to be self-reinforcing and self-fulfilling, they can also come to be regarded as rigid templates for policy-making (Rumley and Gopal, 2007).

From an Australian perspective, much of the eastern portion of the Indian Ocean Region has been characterised as an "arc of instability" (Rumley, Forbes and Griffin, 2006). This arc of instability, which is located within Australia's region of primary strategic interest (ROPSI), has emerged since the end of the Cold War in response to globalisation, the history of colonialism, problems of social and political viability, poor governance, economic instability, aid dependency, ethnic tension and religious fundamentalism.

To a degree, increasing regional instability has been exacerbated by the impact of extra-regional states. During the Cold War period, for example, regional competition for power and influence was especially evident since the Indian Ocean maritime access and basing. In the post-Cold War period, however, the power projection role of extra-regional states, especially China, France and the United States, regional nuclear proliferation and the evolution of sea-based deterrence as well as the emergence of "asymmetric threats" are all additional ingredients of the relatively unstable contemporary maritime security environment (Prabhakar, 2005, 82-3).

In terms of non-conventional threats, for example, after a detailed risk assessment, it has been concluded that a maritime terrorist threat against an Australian vessel or port is a real possibility. It was suggested that one of the reasons why this might eventuate is due to "Australia's close alliance with the United States". However, despite the likelihood of such an occurrence, it seems that the international naval community is still largely unfamiliar with maritime terrorist attacks (Raymond, 2005).

The differential regional impact of globalisation processes has exacerbated levels of Indian Ocean poverty in a region already characterised as the "heart of the Third World". Increasing regional economic security threats in turn create an environment likely to facilitate the growth of non-conventional threats such as civil unrest, terrorism and unregulated population movements. The lack of any robust regional regime of maritime security cooperation ensures that this environment will inevitably lead to an increase in environmental security threats. Such developments potentially threaten the long-term sustainability of the Indian Ocean itself.

Such an environment also facilitates increasing regional threats to human security, which might emanate not only from the impact of authoritarian regimes but also from the strategies employed by some states to combat some of the new non-conventional threats. For example, the Commission on Human Security has argued that the current global war on terror is deficient in at least two ways. First, much state-sponsored terrorism is not addressed. Second, some legitimate groups have been labelled as terrorist organisations in order to stifle any opposition to authoritarian government policies (Commission on Human Security, 2003, 23). While the state still remains the central purveyor of security, it often fails to properly discharge its security obligations and at times can even be a threat to its own citizens. As a result, it is argued that attention must now shift from the security of the state to the security of the people - that is, there is a need to shift the focus to human security (Commission on Human Security, 2003, 2).

In chapter 10, Aparajita Biswas assesses the contemporary state of insecurity in the Indian Ocean Region. On the one hand, it seems that the end of the Cold War, the cessation of major conflicts in the region, and the dismantlement of apartheid in

South Africa have collectively paved the way for the current period of democratisation, regional cooperation and relative peace. On the other hand, however, despite these changes in international and regional relations, Indian Ocean security threats have not been totally eliminated but are reasserting themselves in a fundamentally different form. In particular, in the 21st century Indian Ocean security environment, non-traditional security threats such as environmental degradation, mass population movements, small arms proliferation and drug smuggling have all assumed increasing prominence. The chapter principally addresses the latter problems of small arms and drug trafficking as well as the symbiotic relationship that exists between narco-traders and arms dealers. The author focuses her analysis on a few South Asian and Southern African Countries in the Indian Ocean Region, where the situation of small arms and drug trafficking is the most alarming. Biswas discusses the possibilities for greater cooperation between regional organisations such as the South Asian Association for Regional Co-operation (SAARC) and the Southern African Development Community (SADC), to counter this illicit network in the Indian Ocean Region. She concludes that timing is propitious for a rigorous re-examination of the scope of the problem and argues that focused empirical research could also make a useful contribution to the policy debate.

Vijay Sakhuja in chapter 11 argues that the threat of terrorism has placed special emphasis on trade and transportation security in the Indian Ocean Region and suggests that containers are potentially a significant maritime security threat. Indeed, containers have emerged as the classical Trojan Horses of maritime security. Sakhuja traces the history of container trade and stresses the vulnerability of container trade to terrorism, highlighting technological solutions for its safety and security. He also examines the competing requirements of maritime trade and security and discusses South Asian initiatives with regard to the Container Security Initiative (CSI). Sakhuja notes that, although the South Asian container trade is growing both in terms of volume and value, only one South Asian seaport will soon be CSI-compliant, however.

In Sakhuja's view, even though governments and industries recognise the problem, there is still a need to find new ways to respond to the threat of containers being used as tools for facilitating non-conventional security threats and for illegal activities. For example, there is a need to install equipment to scan containers to examine hidden compartments and inspect the contents without unloading the container. To enable a safe system of commerce, a comprehensive and credible approach to security is essential. The consequences of failure to detect contraband can have devastating security and revenue repercussions and have the potential to slow down the flow of commerce. The author concludes that the remedy lies in stringent and comprehensive detection measures to effectively secure regional maritime trade.

In chapter 12, Mohammed Nizam Basiron and Mokhzani Zubir argue that environmental security is a broad term often used to denote a complex relationship between the environment and all levels of human society. This relationship concerns the impact of humankind on the environment and vice versa and the impacts that these interactions have on human security. The relationship can be observed throughout the world, but is more apparent in the Indian Ocean Region because of the population in the area which has doubled to two billion from 1950 to 1998; its large fishery which includes fifteen million active fishermen (out of the world total of thirty five million); and the relative economic underdevelopment of a large part of the region which means that there is strong dependence on agriculture and fisheries for sustenance.

Basiron and Zubir argue that this situation suggests that any decline in the quality of the Indian Ocean environments, be it in terms of water quality, fisheries resources or marine ecosystems and habitats, could adversely affect the sustenance and livelihood of coastal communities. The authors indicate that there is evidence to suggest that this is happening in the Indian Ocean and explore some of the issues that influence environmental security in the Indian Ocean Region and their implications for environmental management in the area. They conclude that collective regional action is essential in order to preserve the ecological integrity of the Indian Ocean environment which, in turn, will impact directly on the nature of regional human security.

Joshua Ho in chapter 13 describes the increasing use of the major SLOCs in Southeast Asia and considers some of the principal security threats - in particular, piracy and maritime terrorism - to this maritime trade. The role of criminal syndicates in piracy attacks is seen as having a significant impact on shipping costs. Apart from terrorist attacks on shipping, the regional hub ports are viewed as potentially important targets. The author provides a detailed consideration of the various individual, bilateral and multilateral initiatives that have been adopted in order to deal with these threats. While individual, bilateral and multilateral initiatives have been undertaken by the three regional states - Indonesia, Malaysia and Singapore - since the end of the Cold War, multilateral initiatives involving extra-regional states have been mainly driven by the United States. However, the unilateral involvement of extra-regional states in maintaining regional maritime security can exacerbate sensitivities over national sovereignty. Nonetheless, Ho argues that there is still a need to move towards a more cooperative regime involving both the littoral states as well as other stakeholders to enhance the security of the sea lanes since the threats are transnational in nature. The author concludes that, as countries in the region share significant maritime interests, the topic of maritime security needs to remain high on the regional agenda in order to create and maintain a stable maritime environment. The creation of such a stable maritime environment in turn will advance the building of an ASEAN Security Community.

Challenges and Prospects for Maritime Cooperation

The changed maritime security environment of the 21st century has resulted in the emergence of various proposals aimed at meeting new security threats. Many scholars and policy-makers agree that there is now an urgent need in the current global maritime security environment for some form of (perhaps idealist) "regime" or "maritime coalition" to effectively deal with maritime security threats (for example, Bateman, 2005; Kaneda, 2005). In the Indian Ocean Region, it has been suggested that considerable regional scope exists for the development of cooperative maritime security, despite competitive pressures to the contrary (Prabhakar, 2005, 102). While much of the framework for global maritime security has been established through the International Maritime Organisation (IMO), several other organizations have a key role, including the OECD, the World Customs Organisation, the International Labor Organisation and the International Standards Organisation.

International concern has been expressed over the need to combat the spread of weapons technology by sea-based transfer (Prabhakar, 2005, 106). The United States has subsequently implemented its "Proliferation Security Initiative" (PSI), calling upon "like-minded states" to join the programme of ship-boarding and information sharing, with the aim of combating the proliferation of nuclear weapons among so-called "rogue states". The PSI was announced on May 31, 2003 by President Bush as a counter-proliferation programme geared to the suppression of the traffic in unconventional weapons (often called "weapons of mass destruction" or "WMDs"). In September 2003, the 11 states then participating in PSI meetings (the United States, Australia, France, Germany, Italy, Japan, the Netherlands, Poland, Portugal, Spain, and the United Kingdom) set out "Interdiction Principles for the Proliferation Security Initiative" (PSI Principles).

Many commentators have been at pains to identify both the strengths and weaknesses of the PSI, usually on the basis of their particular ideological orientation (Table 1). For example, one such orientation derives from the perceived neoconservative cartographic anxieties of the Bush administration which is seen to underpin the literal text of the PSI. Concerns from this perspective are over the manner in which the United States might be forced to utilise the PSI in practice not only to blur jurisdictional boundaries between the ocean zones, but to seriously undermine the customary international law of the sea. It is in this context that we need to locate Malaysia's growing concern about the possibility of the PSI, wittingly or unwittingly, in undermining its sovereign rights over territorial waters (Rumley and Chaturvedi, 2006).

Table 1 :
Evaluating the Proliferation Security Initiative

Criticisms	Strengths
Inherently unfair	Increases the degree of security capability
Denial of exisiting treaties/conventions	Some states have reneged on non-proliferation
Self-interest of a small group of states	Widespread international support
Definition of terms	Encourages the development of best practice
Loose arrangement	Allows for flexible levels of state cooperation
May hamper legitimate trade	Encourages more secure use of the seas
Potential violation of state sovereignty	Concern over non-state proliferation
Will increase inter-state conflict	Very good survival chances
Scope limited to maritime interdiction	Useful in enhancing overall maritime security
Nature and outcomes of operationalisation	Directed at "rogue states"
Lack of public accountability	Preemptive/Preventive WMD policy

Sources: Gahlaut, 2005; Ho and Raymond 2005; Rumley and Chaturvedi, 2006;Valencia, 2005

Yet others see the PSI as increasing the level of maritime security capability given its widespread international support. The key objective of the PSI Principles was to ensure more coordination in impeding and stopping ocean shipments of WMDs, delivery systems, and related materials "to and from states and non-state actors of proliferation concern". The PSI Principles included a system of interdicting suspect vessels in any part of the world ocean. Despite the criticisms that have been made of PSI, the confirmed North Korean nuclear test of October 2006 is likely to galvanise regional and extra-regional support for the application of some form of PSI which is sanctioned by the United Nations.

In chapter 14, Mat Taib discusses the manner in which transnational criminal activities around the "northern approaches" to the Straits of Malacca have posed a significant threat to its security. This is due principally to the nature of the region's geopolitical environment that has facilitated a longstanding illegal trade in drugs, firearms and human smuggling. Taib's principal aim in this discussion is to argue for a greater and much more effective process of inter-state cooperation at various scales in order to minimise the security threat to one of the world's most important SLOCs. Within the region, despite the presence of ASEAN cooperative security mechanisms against non-state sponsored crimes, genuine cooperation is still lacking. Furthermore, not all regional states are members of ASEAN and thus there are limitations to the existing cooperative process. As a result, the current level of cooperation among regional states is simply no match for the established transnational narcotic and firearms cartels.

Taib suggests that efforts to tackle transnational criminal activities should begin with the creation of an effective national criminal enforcement capability and the harmonisation and updating of national laws. Regional law enforcement capabilities need to be strengthened by increasing trans-boundary cooperation, more importantly in the maritime dimension. Extra-regional state involvement needs to be especially sensitive to sovereignty issues, however. The author argues that India needs to secure its part of the region. He concludes that more research is needed on the politico-diplomatic complexities of these issues before a comprehensive security framework can be formulated.

Manoj Gupta in chapter 15 explores how new post-Cold War legal regimes and geopolitical imperatives provide for freedom of navigation and sea lane security in the Indian Ocean. He argues that, from a geopolitical perspective, effective forms of ocean management to address SLOC security have been rendered more complex by the increased number of nation-states and their differing interpretations of the UN Laws of the Sea. The major maritime powers benefit from the current arrangements or from the absence of them. Furthermore, the globalised economic ideology of states is placing heavy reliance on the market and the role of private enterprise and therefore states are unreceptive to regulatory mechanisms for the oceans. However, UNCLOS III marked an important shift in the balance between 'control and regulation' and the 'freedom of navigation on the high seas'. SLOC security has witnessed a basic change in the attitudes and thinking of navies and coast guards. Gupta argues that the beginnings of cooperative approaches are evident, but comprehensive sea lane security requires a shift from governmental unilateral and bilateral approaches to system-wide arrangements that consider the supply chain in totality. This requires active participation of the logistics and transport industry and a clear mandate to a single agency in government to secure the entire supply chain. The author suggests three sets of guidelines - for collaboration and coordination among coastal states and for cooperation between coastal and flag states - which might form the basis of new SLOC security arrangements in the Indian Ocean.

In chapter 16, Robert Beckman argues that it is possible to establish joint cooperation zones involving foreign patrol vessels in order to enhance the security of sea lanes of communication and combat illegal activities. He examines the establishment of joint cooperation zones in areas outside the sovereignty of the coastal state such as in exclusive economic zones and argues that it is relatively simple to establish joint cooperation zones in such areas. He also examines the establishment of joint cooperation zones in areas subject to the sovereignty of coastal states such as the territorial sea. Beckman argues that it is possible to establish a joint cooperation zone in such areas without undermining the sovereignty of the littoral states if the joint cooperative zone arrangement is established in accordance with nine basic principles. The author considers the question of whether it would be possible to establish a joint cooperation zone in the Straits of Malacca and Singapore and suggests that this could involve joint patrols by foreign patrol vessels in a manner that does not undermine the sovereignty of the littoral states. Such a "burden sharing" agreement would not only enhance the safety and security of the sea lane, but the littoral state would also benefit since it would receive assistance in fulfilling its international obligations without any infringement of its sovereignty. The international community would also benefit because it would enhance the safety of vessels exercising transit passage through one of the most important routes used for international navigation.

Sam Bateman in chapter 17 argues that an integrated approach to ocean management applies as much to maritime law enforcement as it does to other areas of oceans management. As he notes, this reflects the principle enunciated in the Preamble to the 1982 UN Convention on the Law of the Sea (UNCLOS) that has been translated into the well-established concept of integrated oceans management to be adopted both at the national and regional levels. This became an explicit principle, for example, in the Seoul Ocean Declaration adopted for the Pacific Ocean by the Asia Pacific Economic Cooperation (APEC) Ocean-related Ministers in Seoul in 2002. Bateman argues that implementing a multiple use approach to maritime law enforcement that comprehends the full range of illegal activities at sea and provides security against maritime threats requires both cooperation and good information. It employs an integrated approach to maintaining maritime law and order based on agreed international regimes, cooperation, situational awareness, and information sharing.

Of the many institutions involved in comprehensive maritime security, coast guards and port security have an important role to play, but it is essential that their activities are coordinated in a way that achieves the most efficient outcomes and avoids duplication of effort. The author reviews the emergence of coast guard operations in eight Indian Ocean states and discusses the extent of their cooperation. He concludes that there is considerable scope for greater cooperation among all the agencies involved in maritime law enforcement both at the national and regional levels.

Conclusion

It has been argued that a global maritime security agreement is potentially a cost-effective means for restraining US hegemonic ambitions and unilateralism in maritime trade (Stasinopoulis, 2003). Despite the various attempts at initiating such an agreement, it seems that a widely accepted maritime security regime, however, still appears to be a long way off (Bateman, 2005, 260). On the one hand, ReCAAP entered into force in September 2006, while, on the other hand, at the same date, key Indian Ocean regional states were still to ratify the 1988 IMO Convention and Protocol on the Suppression of Unlawful Acts Against the Safety of Maritime Navigation. As has been argued, all states need to ratify the Convention and Protocol in order to signal a clear intent to deal effectively with maritime security issues (Raymond, 2005, 207).

The reasons for the lack of an Indian Ocean Regional Maritime Security Regime (IORMASTER) are complex, but, among other things, they involve a combination of:
- existing regional geopolitical issues
- sovereignty sensitivities of regional states
- insufficient resource capacity
- an inability to view problems holistically
- concern over potential US unilateralism, and
- a lack of individual and collective political will on the part of all states with an interest in maximising regional maritime security.

Securing Indian Ocean sea lanes of communication requires that all of these issues be fully and properly addressed in appropriate regional forums.

References

Amer, R. (1998), 'Towards a declaration on "navigational rights" in the sea-lanes of the Asia-Pacific', *Contemporary Southeast Asia*, Vol. 20 (1), pp. 88-102.

Asia and World Forum *et al* (1980), Forum on the Pacific Basin: Growth, Security and Community, Taipei.

Australian Parliament (2004), *Australia's Maritime Strategy*, Canberra: Joint Standing Committee on Foreign Affairs, Defence and Trade.

Babbage, R. and Bateman, S., eds. (1993), *Maritime Change: Issues for Asia*, St Leonards, NSW: Allen and Unwin in association with the Royal Australian Navy and Australian Defence Industries Ltd.

Banlaoi, R. C. (2005), 'Maritime regional security outlook for Southeast Asia', in Ho, J. and Raymond, C. Z., eds. (2005), *The Best of Times, The Worst of Times: Maritime Security in the Asia-Pacific,* Singapore, Institute of Defence and Strategic Studies: World Scientific, Publishing, pp. 59-79

Barnett, R. W. (2005), 'Technology and naval blockade: past impact and future prospects', *Naval War College Review*, Vol 58 (3), pp. 87-98.

Bateman, S. (2005), 'Maritime "regime" building', in Ho and Raymond, *op. cit.*, pp. 259-275.

Bateman, S. and Bates, S., eds. (1996), *Calming the Waters: Initiatives for Asia Pacific Maritime Cooperation*, Canberra: Strategic and Defence Studies Centre, Research School of Pacific and Asian Studies, Australian National University.

Brzezinski, Z. (1997), *The Grand Chessboard: American Primacy and its Geostrategic Imperatives*, New York: Basic Books.

Bruce, R. H., ed. (1990), *Indian Ocean Navies: Technological Advances and Regional Security*, Perth, WA: Centre for Indian Ocean Regional Studies, Curtin University of Technology.

Commission on Human Security (2003), *Human Security Now*, New York.

Dorman, A., Smith, M. L. and Uttley, R. H., eds. (1999), *The Changing Face of Maritime Power*, Basingstoke, Hampshire: Macmillan Press.

Findlay, T. (1988), 'North Pacific confidence-building: the Helsinki/Stockholm model', *Working Paper*, No. 44, Canberra, Peace Research Centre, Australian National University.

Flanagan, S. J., Frost, E. L. and Kugler, R. L. (2001), *Challenges of the Global Century: Report of the Project on Globalization and National Security*, Washington DC, National Defense University.

Forbes, V. L. (2001), *Conflict and Cooperation in Managing Maritime Space in Semi-Enclosed Seas*, Singapore: Singapore University Press.

Gahlaut, S. (2005), 'Political implications of the Proliferation Security Initiative', in Ho and Raymond, *op. cit.,* pp. 229-250.

Goldblat, J., ed. (1992), *Maritime Security: The Building of Confidence,* New York: United Nations.

Grey, R. W. (1993), 'A proposal for cooperation in maritime security in Southeast Asia' *Working Paper*, No. 274, Canberra, Strategic and Defence Studies Centre, Australian National University.

Harrald, J. R. (2005), 'Sea trade and security: an assessment of the post-9/11 reaction', *Journal of International Affairs*, Vol. 59 (1), pp. 157-178.

Ho, J. and Raymond, C. Z., eds. (2005), *The Best of Times, The Worst of Times: Maritime Security in the Asia-Pacific*, Singapore, Institute of Defence and Strategic Studies: World Scientific Publishing.

International Maritime Organisation (IMO), web site: http://www.imo.org/conventions

Kaneda, H. (2005), 'Regional maritime security outlook: Northeast Asia', in Ho and Raymond, *op. cit.,* pp. 33-58.

Kennedy, P. M. (1983), *The Rise and Fall of British Naval Mastery*, London: Macmillan.

Mack, A., ed. (1993), *A Peaceful Ocean? Maritime Security in the Pacific in the Post-Cold War Era*, St. Leonards, NSW: Allen & Unwin in association with Dept. of International Relations, Australian National University and the Institute of Strategic and International Studies, Kuala Lumpur.

Mahan, A. T. (1890), *The Influence of Sea-Power on History, 1660-1783*, Boston: Little Brown.

Menon, R. (1998), *Maritime Strategy and Continental Wars*, London: Frank Cass.

Paik, J-H. (2005), 'Maritime security in East Asia: major issues and regional responses', *Journal of International and Area Studies*, Vol. 12 (2), pp. 15-29.

Prabhakar, W. L. S. (2005), 'Regional maritime dynamics in Southern Asia in the 21st century', in Ho and Raymond, *op. cit.*, pp. 81-108.

Raymond, C. Z. (2005), 'Maritime terrorism, a risk assessment: the Australian example', in Ho and Raymond, *op. cit.*, pp. 179-212.

Reeve, J. (2001), 'Maritime strategy and defence of the archipelagic inner arc', *Working Paper*, No. 5, Canberra: Royal Australian Navy, Sea Power Centre.

Ricklefs, M. (2003), 'The future of Indonesia', *History Today*, Vol. 53 (12), pp. 46-53.

Roach, J. A. (2005), 'Enhancing maritime security in the Straits of Malacca and Singapore', *Journal of International Affairs*, Vol. 59 (1), pp. 97-116.

Roy-Chaudhury, R. (2000), *India's Maritime Security*, New Delhi: Knowledge World in Association with Institute for Defence Studies and Analyses.

Rumley, D. and Chaturvedi, S., eds. (2004), *Geopolitical Orientations, Regionalism and Security in the Indian Ocean*, New Delhi: South Asian Publishers.

Rumley, D. and Chaturvedi, S., eds. (2005), *Energy Security and the Indian Ocean Region*, New Delhi: South Asian Publishers.

Rumley, D. and Chaturvedi, S. (2006), 'Energy Security and the Indian Ocean Region: Constructing an Indian Ocean Energy Community', a paper given to the 20th World Congress of the International Political Science Association, Fukuoka, July.

Rumley, D., Forbes, V. L. and Griffin, C., eds. (2006), *Australia's Arc of Instability: The Political and Cultural Dynamics of Regional Security*, Dordrecht: Springer.

Rumley, D. and Gopal, D. (2007), 'Globalisation and the new security agenda: developing a regional paradigm', in D. Rumley and D. Gopal, eds. (2007), *Regional Security of Australia and India in the Age of Globalisation*, New Delhi: Shipra.

Singh, J., ed. (1993), *Maritime Security*, New Delhi: Institute for Defence Studies and Analyses.

Stasinopoulos, D. (2003), 'Maritime security - the need for a global agreement', *Maritime Economics and Logistics*, Vol. 5 (3), pp. 311-320.

Stevens, D., ed. (1998), *Maritime Power in the Twentieth Century: The Australian Experience*, St Leonards, NSW: Allen and Unwin.

Till, G. (2004), *Sea Power: A Guide for the Twenty-First Century*, London: Frank Cass.

Valencia, M. (2005), ' The Proliferation Security Initiative: Making Waves in Asia', *Adelphi Paper 376*, IISS: London.

Wilson, D. and Sherwood, D., eds. (2000), *Oceans Governance and Maritime Strategy,* St. Leonards, NSW: Allen and Unwin.

Young, C. (2005), 'Balancing maritime security and freedom of navigation on the high seas: a study of the multilateral negotiating process in action', *University of Queensland Law Journal*, Vol. 24 (2), pp. 355-414.

THE GEOSTRATEGIC IMPORTANCE

OF THE INDIAN OCEAN

Chapter 2

Settlers, Seaways and Subsystems of the Indian Ocean

Jean Houbert

Introduction

The Indian Ocean is an integral part of the global body of seawater that covers over two-thirds of the surface of the misnamed planet earth. Europeans have dominated *the global* ocean throughout modern history. Hegemony over the sea was translated into the European colonisation of virtually all the littoral parts of the planet. It is important here to make a clear categorical distinction between the two kinds of colonies that the Europeans established. The first, and by far the most important kind of European colonies, was a settler-colony. Originally, in Greek antiquity, the word colony meant a group of settlers that left their parent city-state to settle in an uninhabited land and form a new city-state. Cultural and sentimental links were kept between the settlers and the parent city-state, but no political domination was involved. This kind of colony is seldom found in its pure form in modern history. In modern times, the parent-state has endeavoured to keep its rule over the colony, and, more significantly, native inhabitants already occupied the land where the colonists settled.

The settler-native distinction is to a degree "constructivist". There were, of course, differences of functions, of social class, of religion, of gender among the settlers. The natives never formed a homogenous group. Yet the settlers had no doubt who was and who was not of their group, who was "the other", when they came in contact with the natives. The natives rapidly came to think of themselves as such. Racial/ethnic difference trumped all other distinctions. The contact between settler and native was experienced by the former, not as a "clash of civilisations", but as a clash between the civilisation and barbarism.

Settler-colonisation in modern history has involved a double domination: that of the parent-state over the colony and that of the settlers over the natives. The interests of parent-state and settlers were not always in harmony; they frequently diverged over the treatment of the natives. The European parent-state often wanted to protect the natives by insisting that part of the land of the colony be reserved for the natives. There was also on the part of the parent-state the desire to preserve the native way of life, although there was tension between this and the European ideology of progress. At the limit, *a reservation* was like a human-zoo for "endangered species".

Settler-colonisation was as much the result of the push from Europe than of the pull of the new worlds. The settlers that went overseas were seldom the best element of Europe. Indeed, the settlers were frequently those that Europe rejected: criminals, prostitutes, orphans, bankrupts, unemployables, rebels, failed revolutionaries, persecuted religious minorities, and so on. These and many other kinds of misfits were transported or shipped themselves out to the settler colonies. Yet the settlers, far away from Europe, often saw themselves as living on the frontier of civilisation advancing against barbarism. There is in the Judeo-Christian tradition an element - more frequently present in some variants of Protestantism than in Catholicism - that perceives itself as the "Chosen People of God". In the circumstances of settler-colonisation this was often interpreted as a God-given entitlement to appropriate the land of the natives and rule over them.

A successful settler-colony in modern times is one where the settlers have appropriated the land and have found an irreversible solution to "the native problem". Decolonisation in a successful settler-colony means the transfer of political power and sovereignty from the parent-state to the settlers. This comes about through the successful use of force by the settlers against the parent-state, as was the case in the United States, or by the parent-state granting self-determination peacefully to the settlers, as was the case in Australia. The end of the rule of the parent-state does not necessarily bring to an end the domination of the settlers over the natives. On the contrary, with the restraining influence of the parent-state gone, the powerful settlers felt free to find a final solution for "the native problem". Solutions have ranged from expulsion to forceful assimilation; at the limit, the settlers resorted to genocide to solve "the native problem". An unsuccessful solution of "the native problem" can result in the failure of a settler-state.

Other peoples and civilisations have practised settler-colonisations in history. However, it is successful European settler-colonisation, closely linked with sea power, which has changed the course of modern history and radically transformed the world. Not least because European settler-colonisation gave birth to the United States, the hegemonic thalasocracy of the present day and the Superpower of the unipolar international system. No doubt that the world has benefited from the successful European settler-colonisation of large parts of the planet; but humanity has also paid a heavy price: "ethnic cleansing", genocide and ethnocide on a scale that the world had never known before.

In time, the word colony came to be used also for a different phenomenon altogether. In this second kind of colony, a few European administrators, soldiers and entrepreneurs, backed by the sea power of the distant metropolis, ruled over the natives with the collaboration of local elites. Notwithstanding Kipling's *East is East* and *West is West*, the "clash of civilisations" between rulers and ruled was perceived by the former through an ideology of progress whereby the natives would be transformed in the image of Europe. The native elites, while retaining elements of the traditional cultures, were keen to acquire the institutions - and, in particular, the modern state and the capitalist mode of production - that had made Europe rich and powerful. Once these institutions were well implanted, this kind of *non-settler-colony* became redundant and could be decolonised. Here decolonisation means the transfer of political power to a native elite within the territorial boundaries of the colony that were drawn by the erstwhile European colonisers. On the whole, this was a peaceful process. Cases where large-scale violence was involved were few. Decolonisation of non-settler-colonies in turn has contributed to some extent to *deracialisation* in the international system.

The littoral of the Indian Ocean was the domain par excellence of the second kind of colony, the non-settler-colony. The first kind, the settler-colony, was implanted principally in the western hemisphere. The Indian Ocean, however, is virtually landlocked, opening freely only to the south. This geographical feature has given the chokepoints of the seaways in and out of the Indian Ocean considerable strategic and commercial significance. Right from the time that the European navigators first discovered that the Indian Ocean was not a gigantic lake but a part of the global ocean, the Sea Powers have been keen to control its outlets. In the offing of the all-important seaways linking the Indian Ocean and the rest of the global ocean, three European settler-colonies were implanted and grew up to become settler-states in the international system. The oldest of the three by far, South Africa, grew up from the European settler-colony implanted at the tip of Africa on the Cape route, the first of the seaways to the Indian Ocean that the European navigators discovered. The second was Australia, established as a British penal settlement, but with an eye of the thalasocrasy on Malacca and the other seaways through the Indonesian islands connecting the Indian Ocean and the Pacific. The third, Israel, in the offing of the Suez route, is much more recent. Indeed, Israel is unique as a settler-state that came into being when the decolonisation of non-settler-colonies in the Indian Ocean had already started.

The demographic balance between settlers and natives stands out among the factors determining the fortunes of these three settler-states of the Indian Ocean. Land was the essential stake in the confrontation between settlers and natives everywhere. The outcome of that confrontation depended on the number of settlers relative to the number of natives; and this could be affected by birth rates and immigration but also by the policy adopted by the settlers toward the natives. The policy in turn was conditioned by whether or not the settlers needed the natives for labour. When the settlers were numerous and militarily strong while the natives were few and not resilient, in particular, when they had no resistance to the diseases introduced by the settlers, and when, furthermore, the settlers wanted the land but did not require the labour of the natives, the fate of the latter was sealed.

The structure of the international system also remains a crucial element in the fortunes of the settler-states. Within that structure the relations of the settler-states with the dominant power at sea is decisive. By structure of the international system is meant the configuration of the balance of power in the world. Originally, the international system was exclusively European. In modern history, European civilisation has been subdivided into nation-states co-existing in anarchy - in the sense of the absence of a government over and above the sovereign states. Through the dynamic of European colonisation-decolonisation this international system became global. After *The New Thirty Year War* (1914-1945) the international system ceased to be Euro-centric and polarised on the two Superpowers of the Cold War, the USA and the USSR. Bipolar, the Cold War international system was also *heterogeneous* in that the political regimes and ideologies of the two Superpowers were mutually antithetical. With the disappearance of the USSR and the end of the Cold War the international system became unipolar; and there is a tendency now for the sole Superpower, the United States, to translate its hegemony into the *homogeneity* of the international system.

The international system is global, but one of its characteristics is the presence of subordinate systems (sub-systems) in different parts of the world. The failure of a settler-state to find an irreversible solution to "the native problem" can lead to the emergence of an international subordinate system as the internal conflict between settlers and natives interacts with the global international system. This is the more likely when the internal settler-native conflict spills over in a region, with neighbouring states identifying with and taking sides in the conflict.

Of the three settler-states of the Indian Ocean, Australia is the only one that has been very successful in its solution of "the native problem". Timing, geographical location, above all demographic circumstances, were favourable. The international sub-system(s) in which Australia partakes today has a lot to do with the seaways, but not at all with "the native problem", for that has virtually ceased to exist. South Africa is at the other extreme: the solution to "the native problem" was never viable and has been reversed. Indeed, South Africa may well now have a "settler problem". Israel, so far, has been more successful in its solution of "the native problem" than was South Africa, but the settler-state here has nothing like the security and legitimacy of Australia.

Following is a brief case-by-case analysis of the three settler-states in the offing of the seaways of the Indian Ocean.

South Africa

The demographic balance between settlers and natives has always been heavily against the European settlers in South Africa. The settlers wanted the land but also the labour of the natives to work the land and later the mines and the industries. From very early on the settlers shunned manual labour and relied on the natives. Indeed, it became almost a part of the identity and self-valuation of European settlers not to do manual work, and especially not alongside natives. This made for the exploitation of the natives, but it ruled out genocide as a means of redressing the demographic balance. What developed instead was partition of the land on a very unequal basis with the settlers annexing the bulk of South Africa and forcing the much more numerous natives into what were called reservations. Overall sovereignty in the country remained with settlers' South Africa, however, even after the partition was systematised under apartheid. The policy of forcing the natives to live in Bantustans, the renamed reservations, and work in settlers' South Africa as pass-carrying guest workers, was dysfunctional for the economy. Attitude to work also underpinned job-reservations before and during the apartheid era that created a dual work force of settlers with a First World standard and natives at the level of Third World. Apartheid South Africa ultimately failed not least because it was a system that used state power based on race to prevent the development of the normal social features of a capitalist economy, such as class mobility across racial lines.

The interests of the settlers in South Africa and the interests of their parent-state often conflicted over the question of the natives. This was the more marked in South Africa after the original settlers lost their Dutch parent-state and a new group of settlers came in with the new parent-state, Britain. The Dutch were great sailors and their country had been an outstanding sea power and coloniser. Size and location, however, made Holland vulnerable to the land power of France in Europe. Britain, the rising sea power, captured the Dutch colonies overseas to pre-empt them falling to France during the Napoleonic wars. Indonesia was later returned to the Dutch but Britain kept the strategically located colony on the Cape route. The Dutch settlers resented British interference with the way they treated "their" natives and this led many of them to move inland out of reach of British law. The Great Trek, ideologised as a repeat of the Old Testament's "Chosen People" seeking the

"Promised Land" of milk and honey, brought the Dutch settlers into more conflict with the natives over the land and eventually to the founding of two new settler-states, Transvaal and Orange Free State, independent of Britain. The discovery of diamonds and gold inland attracted British interests. It was with somedifficulties, however, that Britain, although the foremost imperial power, and despite hegemonic power at sea, defeated the Boers. The unification in a new Dominion of South Africa of the Boer settler-states and the British colonies, Natal and the Cape, did not eliminate the ethnic difference between Dutch and British settlers. This ethnic rivalry between the two groups of settlers, however, did not much benefit the natives as political power was transferred from Britain to the settlers, who closed rank on a racial basis. The term ethnocracy has been suggested for the kind of political system where democracy obtains for the settlers but excludes the natives. In South Africa, all the European settlers had the vote on the basis of race but the ethnic distinction between Boer and British remained. Perhaps raceocracy would be a better term here than ethnocracy.

Settlers' South Africa was not considered to be an anomaly so long as the international system remained Euro-centric and colonial rule was regarded as normal. The change came with the Cold War and the polarisation of the international system between the two Superpowers and the rapid decolonisation of non-settler-colonies. One of the stakes in the conflict between the two Superpowers was the new state coming out of the decolonisation of the colonial-empires of Europe. All of the European colonial Powers were the allies of the United States in the Cold War and were dependent on that Superpower for their military security in Europe and also for a time for their economic survival. Washington was therefore well placed to exercise considerable leverage on its European allies to get them to transfer power to the Westernised native elites in the colonies, lest these be displaced by more radical elements that might then turn to the Soviet Union. In this new international system, where the new recently-decolonised states became the majority in the United Nations, apartheid South Africa became a *pariah*.

The Cape route, however, remained important for the West during the Cold War and some of the minerals found in South Africa were irreplaceable. The Cape route was important commercially as well as militarily. The Suez route was vulnerable in local wars between Israel and the Arabs and would have been even more so in a general war. The giant tankers taking oil to the West could not use Suez and the economies of scale in maritime transport made it as cheap to go by the Cape than to use smaller tankers via Suez and paying the canal dues. The loaded northbound tankers sail close to South Africa to avoid the Roaring Forties and the enormous seas of the Southern Ocean that can slow down or even seriously damage the giant ships. There is also a commercial reason for keeping as near as possible to the southern hemisphere which comes for some months under the lighter winter load-line. The International Maritime Conference Organisation, however, had agreed that, provided that the ships sail close to South Africa, the tankers could load up to the tropical mark when they leave the Gulf. South Africa provided - and it still does today - special repair facilities, as well as fresh food and mail, while crew replacements could also be flown out to the tankers as they rounded the Cape. The nearby communication networks plot the ships a long way off. The naval and air power of the Republic of apartheid covered the ships until they were clear of land into the Atlantic and out of reach of any threats from the land.

In 1955, Britain had transferred the naval base in Simonstown to South Africa in an agreement that provided for: "Southern Africa and the sea routes round Southern Africa must be secured the lines of communication and logistic support in and around Southern Africa must be adequate and securely defended". Later, in 1964, Britain, disapproving of the settlers' policy of apartheid, imposed an arms embargo on South Africa. However, in 1971, the British Law Office took the view that Britain was obliged to supply certain kinds of arms to South Africa for the defence of the sea route. Militarily, in a general war, it would have been advantageous for the West to block Suez and use the Cape route where it was assured of the South African facilities. While the Soviet war ships, without the use of Suez, would have lost their naval short cut to the Indian Ocean and their Far East base. Moreover, the settlers were well entrenched in power in South Africa and were well armed. To drive the settlers out militarily was beyond the means of the natives, short of massive external military assistance; and that, in the Cold War, could come only from the Soviet bloc.

The conflict between the settlers and the natives inside South Africa spilled over into the neighbouring states giving rise to an international subordinate system. For a time, the apartheid state had allies in the Portuguese colonies of Mozambique and Angola and also in UDI Rhodesia. While the natives who challenged apartheid had the support of the newly-decolonised African states, and, in particular, those that came to be called the *Front Line States*. However, the balance of military power was overwhelmingly on the side of the settlers. To the extent that the sub-system of Southern Africa polarised on the Cold War it brought the Soviet Union on the side of the natives. The use of Cubans to intervene in the Southern Africa international sub-system was less risky for Moscow than sending Soviet troops in a region where the West, dominating the seas, and assured of the support of South Africa, was militarily advantageously placed. The West relied on South Africa to repulse the Cubans while pressurising Pretoria to make concessions to the natives, particularly on Rhodesia and on Namibia. As the Cold War drew to an end, the Superpowers supported a deal whereby the Cubans would leave Southern Africa, and Namibia, the "colony" of South Africa, would become independent.

The settlers were not defeated militarily by the natives in South Africa. On the contrary, the apartheid regime negotiated itself out of existence when it was in a position of strength militarily. The arms boycott decreed by the United Nations had had the effect of boosting arms production in South Africa and of joint development of weapons, including nuclear weapons, with Israel. The departure of the Cubans, the disarray and then the collapse of the Soviet Union, left the natives without powerful military allies. The United States, itself a settler-state, and now the sole Superpower, had had reservations about the role of the Communist Party of South Africa in the native organisations opposing apartheid. However, now that the Cuban-Soviet threat in Southern Africa had disappeared, Washington was pleased to see the African National Congress (ANC) and the South African government coming together to negotiate a way out of apartheid. The outcome is that political power has passed from the settlers to the natives. This is irreversible. Economic power, however, remains largely with the settlers. In terms of social status and employment, the principal losers to date have been the petty settlers, in particular, those who occupied lesser positions in the state apparatuses by virtue of race. They have had to make way for natives with better qualifications and political support. The principal winners have been the natives "state-bourgeoisie". "Affirmative Action" and Black Economic Empowerment (BEE) probably will in the long run move the society away from one structured by race to one structured by class. However, for the time being, the settlers still occupy the top of the social pyramid and the natives are at the bottom. The pressure of the global capitalist economy makes for more privatisation and this limits the possibilities for the state to intervene in the economy. In any case, with 40% unemployment among the natives, the government needs all the economic growth that the "private sector" can deliver.

The settlers of European antecedents, given their expertise in several fields of modernity and their control of capital, are, for the time being, more of an asset than a liability for the post-apartheid South Africa. Ethnic cleansing, in the sense of the eradication of the culture(s) associated with the settlers, is improbable. In any case, after centuries of setter-colonisation there is cultural ambiguity in South Africa. There is a substantial number of mixed race in the population who are closer to the settlers culturally than to the natives. Very many of the natives, and in particular the elites, have adopted cultural features that were initially associated with the settlers, such as the English language and variants of Christianity. Above all, the principal institutions of modernity that Europeans have spread around the world - the capitalist mode of production and the modern state - are well rooted in South Africa and will remain even if all the settlers were to leave. The likelihood is that those settlers who do leave would do so voluntarily rather than be forced out by the natives. The Truth and Reconciliation Commission did not deal with the structure of South Africa's society; it was premised on a "truth for amnesty" deal between the ANC and the National Party (NP), the governing party under apartheid. The NP, renamed the New National Party, has now merged itself in the ANC. Demographically, the natural rate of reproduction of the settlers is much lower than that of natives, despite the heavy toll of AIDS among the latter.

The Southern Africa international sub-system that had developed around the settler-native conflict in South Africa has ceased to exist. The priorities of the present regime in South Africa are much less military than were those of the apartheid state. South Africa is the only nuclear-weapon-state that has voluntarily given up its weapons. South Africa nonetheless remains an economic and political giant in Africa. There are a number of "quasi-states" in Africa, that is, states that are sovereign more through membership of the international system than through their own viability. Some of these states might fail altogether and the resulting chaos would provide suitable conditions for anti-Western extremists to thrive. The Cape route retains commercial and strategic importance and The United States is well pleased to see the new South Africa as a factor of stability in the region.

Australia

Australia is by far the largest settler-state outside the Western hemisphere. It is also the settler-state that has been the most successful in solving "the native problem". So successful in fact that it was difficult to think that Australia had ever been anything else but another Europe on the other side of the world. Yet, for some sixty thousand years before the arrival of the First Fleet, natives had inhabited the island-continent. The natives had penetrated and occupied all parts of the country, including Tasmania before rising seawater cut off that island from the mainland some twenty-two thousand years ago. Estimates of the number of natives when the first European settlers arrived in 1788 vary between 300,000 and one million. It took the European settlers less than one hundred years to virtually wipe the natives off the face of the continent. The few natives who have survived, mixed to varying extent with settlers, form a very marginal group that can present no threat of any kind to settlers' Australia.

In a country the size of a continent there was ample room for a small population of natives and for the settlers that even today, after two hundred years of colonisation, are not numerous. Co-existence of natives and settlers, however, was difficult because although Australia is immense, only a fraction of the continent is well watered. The natives were hunter-gatherers, a way of life that required a great deal of space to sustain a small population. Because the country is so dry, the water holes were crucial for the native economy. The kind of land-economy that the settlers adopted, cattle and sheep rearing, also required extensive land, some water, but few workers. The natives were expelled into the waterless areas where they could not survive or were shot and poisoned. The natives were exterminated because the settlers wanted their land but had little need of them for labour. The technology of armaments of the settlers was much superior to that of the natives who could be killed off with little or no risk to the settlers. Ideologically, the settlers, at best, rationalised the elimination of the natives as the inevitable disappearance of "primitives" in contact with civilisation. Darwin, who visited Australia when the cleansing was underway, thought that the chances of survival of the natives were slim. The "survival of the fittest" became part of the credo of European racists. B.D. Moorhead, later Prime Minister of Australia, wrote: " the colonist had come here as white men and were going to put the black man out the lower race must give way before the superior race". The parent-state, Britain, had all along wanted to protect the natives. Thus, Phillip's official instructions were "to endeavour by every means in his power to open an intercourse with the natives and to conciliate their goodwill to live in amity and kindness with them". Later, the Letters Patent establishing the colony of South Australia in 1836 said: "Nothing should affect the rights of the natives in regard to their enjoyment or occupation of the land". Britain voiced some objections to the solution of "the native problem" adopted by the settlers. The Westminster Parliament, however, was far away and London was reluctant to intervene in the affairs of the self-governing colonies. Moreover, industrialising Britain needed more and more wool and meat and the settlers knew best how to produce these in the large-scale use of land in Australia.

Natives were not even counted in the census let alone have the right to vote in Australia until 1971, when most non-settler colonies had been decolonised. In 1997, after Australia's treatment of the natives had been denounced as racist by the United Nations in 1988, the Human Rights and Equal Opportunities Commission released a document entitled Stolen Generations giving details of what amounted to a policy of ethnocide between 1910 and 1970. During this period, the children of the few remaining natives were removed from their mothers and relocated in settler families and orphanages so that they could be brought up as real Australians. The document recommended an official government apology. It was only in August 2000 that the Australian Parliament passed a motion expressing "deep and sincere regret" for the injustices. The government, however, no doubt anxious to stall the white backlash and win over the one million who had voted for the openly racist party of Paula Hanson's One Nation, stopped short of an apology.

In June 1992, the Australian High Court, in the so called "Mabo" decision, said that

Aborigines and Torres Strait Islanders should be able to claim native title to land if they could show a "close and continuing" relationship with the land in question. The ruling was said to be a "landmark" in that it recognised for the first time the existence of land title before the first European settlement in 1788. This overturned the concept of *terra nullius* (land belonging to no-one) that Europeans had used since the Age of Discovery to appropriate the countries of the "uncivilised" around the world. "Mabo" established a new entitlement to land not grounded in statute law. Settlers, and in particular mining interests, protested. The State of Western Australia took the Federal government to the High Court contesting the validity of the Native Title Act. In March 1995, the High Court ruled against Western Australia and in December 1996, in the so-called "Wik" ruling, the High Court ruled that native title might continue to exist on land subject to a pastoral lease. This ruling has caused tensions in many rural communities where the settlers are worried about the legislation's effect on their livelihood.

Mabo and Wik are no doubt great victories for the natives. The recognition of title to property in land, however, cannot undo the transfer of Australia as a whole, as a country, from the natives to the settlers. Leaving aside the fact that there are very few natives left, at best, the natives can get title to property in land within the sovereignty of the settler-state. Sovereignty over the country belongs to the settlers. The High Court of Australia is an institution of the settlers in a sovereign Australia. There never will be an Australia under the sovereignty of the natives. In that sense, the settler-colonisation is irreversible. Belatedly, in the twenty-first century, the Australian authorities, anxious to project abroad a non-racist image after the UN censure, have highlighted the native contribution to Australia's identity. Exhibitions of Aboriginal artwork are put up by the embassies in foreign countries. At the Sydney Olympic games, the Australian teams included athletes with native antecedents. National pride was taken in Cathy Freeman winning a gold medal.

The settlers who had no need for the native labour also preferred to use a small but highly paid labour force of European settlers rather than allow the entry into Australia of cheap labour from Asia. European convict labour was available at first, then the ex-convicts and free settlers, from very early on, were well organised enough to prevent Asiatic immigration. "Rule Britannia, Britannia rules the waves and no more Chinese in New South Wales" was one of the slogans in the 1870s when the White Australia policy was being instituted. Democratic self-government meant that it was important politically to win the settler-workers vote. Australia thus avoided being trapped in the cheap-labour low-production syndrome of underdevelopment. The price of the highly paid workforce, with a more than European standard of living, is that the population of Australia has remained low, and this, with the teeming populations of Asia on its doorstep, is a factor of weakness in international relations.

Australia, modern settler Australia, was born of the sea. It is in the offing of Asia, but it is its continental size, small population of European settlers, and insularity that have been determinant. For a long time it was easier to travel by sea than by land over the enormous distances of the island-continent. The vast majority of Australians now live on the coast in the port-cities. Australia has remained very much oriented to the sea for its economy as well as for its security. Yet, except for yachting, Australians are not seafarers. Unlike the American settlers who from early on were great sailors and challenged the navigation monopoly of the parent-state, the Australians gladly left the carrying trade over the seas as well as the maritime defence of the island-continent to Britain. Alone, the capabilities of the small population of Australia to defend such a large country would not have been credible enough to deter a large enemy bent on raiding or even invading. Allied to the power dominating the seas, the island location facilitated the defence of Australia from the sea. Britain's hegemony over the global ocean provided that defence.

Settlers are usually dependent on their parent-state for security towards the natives, towards rival European states and towards local neighbouring states. In the case of Australia, after the initial implantation of the settlement, the settlers were more than capable of dealing with the natives without help from the parent-state. After Trafalgar, Britain had hegemony over the global ocean for over a century. The Australian settlers were concerned when, in the so-called Second Imperialism, European Powers other than Britain returned to the Indian Ocean and the Pacific, notably France in Indo-China and Germany in New Guinea. However, it was the rise of Japan, the only Asiatic state that succeeded in modernising itself rapidly enough to join the ranks of the Imperialist Powers that transformed the configuration of power in the Asia-Pacific region. The arrival of Japan as a major military power, and in particular with power at sea, heralded a disjuncture in the perceptions of security interests in Britain and Australia. Britain was not displeased that Japan's defeat of Russia in the Far East relieved the pressure of containing the Tsarist's push on the frontiers of British India. The alliance with Japan enabled Britain to concentrate its naval forces in the Atlantic. Australia, however, perceived Japan as the "Yellow Peril" that sooner or later would pose a major military threat. Later, after Britain had abandoned the Japanese alliance, the so-called Singapore policy was meant to reassure Australia. This was largely a bluff: the building of a large naval base in Singapore was supposed to deter Japan by signalling Britain's intention of sending a major fleet to the region in a crisis. It was clear, however, that Britain could not simultaneously confront major enemies at sea in the Atlantic and the Mediterranean and at the same time face Japan in the Pacific. The outcome was the debacle of 1942. Fortunately for Australia, a fellow erstwhile British settler-colony, the United States, was ready to take over from Britain the role of thalasocratic hegemon.

During the Cold War, Australia shared the interests as well as the values of the United States and participated in all of the wars to contain communism in Asia. In the post-Cold War unipolar international system, the strategic doctrine of the United States, as spelled out in Forward from the Sea, in 1994, is one of projecting power from the sea to the land in the littoral of Eurasia-Africa, what MacKinder called the World Island. The traditional distinction between land forces, air forces and sea forces is less important in this essentially sea power strategy. Command of the sea, that Mahan had envisaged being gained through a "decisive battle" in which the enemy's fleet is defeated or through its neutralisation, is now taken for granted. The United States does not have to acquire command, it already rules over the global ocean. The strategy can therefore concentrate on the projection of power in the littoral of the World Island. While during the Cold War confrontation was with another Superpower and the primary theatre and stake was Europe, now it is Asia and confrontation ranges across the whole spectrum of peace, crisis and war. The threats, potential and actual, are perceived as coming from non-state actors as well as states. In balance of power terms, China is included on the list of threats as a possible future challenger for hegemony. However, more urgent threats are perceived as coming from "rogue states", "terrorists", illegal immigrants, the transnational drug trade and failed or failing states. Since 9/11, the dread is that anti-Western terrorists will acquire weapons of mass destruction. Novel methods in the asymmetrical multi-faceted confrontation called for a revolution in military affairs (RMA) involving intelligence, use of information technology, "smart" weapons, unmanned aircrafts, forward bases, pre-positioning equipments, lift-capacity, advanced landing-crafts, and so on.

The virtual elimination of the natives of Australia has precluded the emergence of an international sub-system based on the settler-native conflict. In 1971, Australia, at long last, discarded its White Australia policy; but as the Afghan asylum seekers incident of August 2001 demonstrated, it is prepared to go to great lengths to enforce its strict immigration policy. Bali, in October 2002, indicated that Australia could be a target of transnational terrorism. Australia has plugged-in the American strategy in Asia and Washington is grateful for the part it plays in reinforcing security in the region. When asked if the United States considers Australia to be the "Deputy-Sheriff" in the region, President Bush said: "No, not the Deputy-Sheriff; Australia is the Sheriff". Australia is active in sub-systems in Asia and the Pacific. Indonesia's archipelagic fragility is a factor of concern; but Australia did not hesitate to take the initiative over East Timor in 1999.

Of the three all-important seaways linking the Indian Ocean with the rest of the global ocean, Suez and the Cape were crucial militarily and commercially in the Euro-centric international system. They remain significant in the present system.

With the hegemony of the United States, however, it is Malacca and the straits through the Indonesian islands that are indispensable; they are the principal channels through which America brings its sea power to bear in the littoral of the Indian Ocean. The major part of the oil exported from the Gulf goes through these straits to Japan, the principal Asian ally of America, and to its other allies in the Asia-Pacific region. The straits are vital to Australia for its exports and imports as well as for its military interventions in the Gulf alongside the United States. The blockade of these straits through terrorist activities or state failure would be a major blow to thalasocratic power.

Israel

Israel is a settler-state with several differences. It is unique as a successful settler-state that came into being when the decolonisation of non-settler colonies was under way followed by deracialisation in global politics. The settlers claim to be the original natives that had to leave Palestine in ancient history and have now come back to their original land. The land, the settlers claimed, was "empty" in the *terra nullius* sense; the Arab inhabitants were seen to be merely squatters who had to make way for the return of the authentic natives. The solution to "the native problem" that the settlers adopted - expulsion - was not acceptable to the natives who have tried, unsuccessfully, to reverse it by force. The political parent-state, Britain, was not the main ethnic parent-state of the settlers, who came primarily from Eastern Europe.

With the demise of the Turkish Empire in the First World War, Britain had been keen to acquire its Palestine province as a buffer for the protection of the northern approaches to the then all-important Suez Canal, the lifeline of the British Empire in the Indian Ocean. Largely for the political reason of winning Jewish support in Europe and the United States in the war, the British authorities were receptive to the arguments of the leaders of the Zionist movement who wanted to create a state for the Jews in Palestine. The British leaders, Lord Balfour in particular, also subscribed to the Old Testament ideas of the "Chosen People" of God and the "Promised Land". Three-quarters of a century later, Balfour was echoed by the President of the United States, Jimmy Carter: "I consider this homeland for the Jews to be compatible with the teaching of the Bible, hence ordained by God". Furthermore, a Jewish settler-state created by Britain would be grateful and remain a reliable ally in the defence of the Suez Canal. Theodore Herzl, the founding father of the Zionist movement, had said in characteristic European settler's language: "For Europe, we would constitute a bulwark against Asia down there; we would be the advance post of civilisation against barbarism".

Palestine, however, was not "empty", since Arab-speaking inhabitants had been there all the time. To win Arab support in the campaign against the Turks, Britain had promised that independent Arab states would be created once the Turkish empire had been defeated. The Sykes-Picot agreement between Britain and France to divide between them the Arab part of the Turkish Empire did not help to reconcile the conflicting promises made to the Zionists and the Arabs. Zionists have argued that when Britain got the League of Nations mandate it was committed to make the whole of Palestine a Jewish national home. Mindful of the native Arab Palestinians, Britain stressed that its commitment was for a Jewish national home in Palestine but not of Palestine. In the beginning, the Zionist settlers were only a tiny minority of the population of mandated Palestine. However, as anti-Semitic policies gathered strength in Europe, the number of Jewish settlers in Palestine increased, despite the strong objections of the native Arabs and the efforts of the British authorities to limit the numbers.

As the flood of settlers rose, with the United States adding pressure on Britain to let them in, the usual conflict over the treatment of natives between settlers and their parent-state took an increasingly violent form, with the Zionists adopting terrorist methods. Britain finding Palestine too hot to hold handed the mandate back to the United Nations which then proposed the partition of the country. In Zionist thinking, the whole of Palestine belongs to the Jews. As a temporary measure, however, and as the UN plan provided for a larger share of the country going to the settlers, although they were still considerably fewer in number than the natives, the settlers accepted the partition proposal. The natives, confident in the legitimacy of their rights and of the support of the Arab world, rejected the partition of their country. The neighbouring Arab states identified themselves with the cause of the native Palestinians and moved their armies into the territory to confront the settlers. In the ensuing war, the settlers were victorious and proclaimed the state of Israel in a larger part of Palestine than the UN partition plan had envisaged. The settlers wanted the land of Palestine, but being largely self-sufficient for labour, had little or no need for the natives. During the fighting, subsequently the majority of the natives were expelled into neighbouring Arab states, including into the West Bank, the part of Palestine that Jordan annexed during the war. The settler-state, Israel, was thus born out of superior military force; it has existed and grown since largely through predominant force.

The Israeli victories against the Arab states in the subsequent wars, all rooted in the native-settler conflict over Palestine - Suez, Six Day, Yom Kippur, Lebanon - gave more Arab lands to Israel. The military victories and territorial aggrandisement, however, did not resolve "the native problem". On the contrary, they added to it because with the lands acquired came more native Arabs who could not be integrated into the settler state. The principle on which the Israeli state is based is that of a state exclusively for Jews. The policy of Israel regarding the acquired territories is the so-called "land for peace" policy. This consists of handing back the conquered territories to the Arab states in exchange for recognition of the legitimate right of the settler-state to exist. So far, the policy has been successful with Egypt and with Jordan. Largely for security considerations, Israel has not handed back to Syria the Golan Heights that it took in the Six Day War. "Land for peace", however, has not worked with regard to the core problem, that of the natives of Palestine.

Over the land of Palestine the settlers and the natives have mutually exclusive objectives. The difference is that, after several military defeats of the Arab states that identified with the native Palestinians, at least some of the leaders of the Palestinian natives - those of Fatah not those of Hamas - have come round to accept the partition of their country and would be agreeable to a "two states in one country" solution: a settler-state and a native state. Israel, on the other hand, for security as well as for ideological considerations, is not enthusiastic to see a full sovereign state of native Palestinians alongside the settler-state. If the Zionist claim is valid, that the settlers are in fact the authentic natives of Palestine that have come back home, then that claim is valid not just for a part of the country but for the whole of Palestine. The West bank, above all the whole of Jerusalem, are as much integral parts of *Eretz Israel* as Haifa or Tel Aviv. To accept a native Arab state in Palestine would thus weaken the legitimacy of the Zionist claim, the very ideological foundation of the state of Israel. Furthermore, a native Arab state in Palestine would further encircle Israel with potential enemies. Water supplies from the Jordan River to Israel would also be more difficult to control. More agreeable for Israel would be that the native Palestinians accept a permanent variant of the present Bantustan-like solution, but without intifadas. That is, some political autonomy for Palestinians in Gaza and the West Bank, with blocks of Israeli settlers dispersed among them, linked by military roads and fortified places controlled by the Israeli army.

Envisaged in the first place in connection with the Suez Canal, the settler-state has been dysfunctional to that seaway. Suez, being a manmade canal across Egypt, has a different status under international law to that of natural straits. The Suez Canal is subject to the Constantinople Convention. It is that Convention, and not the Suez Canal Company, as was wrongly argued by Britain and France during the Suez Crisis of 1956, that guarantees freedom of navigation in the canal. Under the Convention, ships of all states can use the canal in peace and in war. Egypt is charged under the Convention to implement the guarantee of freedom of navigation in the canal. Egypt has sovereignty over the territory on which the Suez Canal is built. When Britain was in control of Egypt it stopped the ships of its enemies from using the Suez Canal in wartime. Britain argued that enemies' ships in transit trough the Suez Canal could commit acts of war that would damage the seaway. This could make the Suez Canal unusable and would thus be a breach of the Constantinople Convention. Following the British precedent, Egypt denied Israeli ships the use of the Suez Canal, arguing that, as it was at war with Israel, the ships in transit could do damage to the seaway. Note that Egypt stopped the Israeli ships well before Nasser nationalised the Suez Canal Company in 1956.

Israel, denied the use of the Suez Canal, relied heavily on its port of Eilat at the head of the Gulf of Akaba to gain access to the Red Sea and thus to the Indian Ocean. The Strait of Tiran, at the mouth of the Gulf of Akaba, falls into the territorial seas of Egypt and Saudi Arabia that could close it to Israeli shipping. Guns at Sharm el Sheikh on the edge of Sinai controlled the use of the Strait of Tiran. One of the reasons that led Israel to associate itself with Britain and France in the Suez war of 1956 against Egypt was to gain control of Sharm el Sheikh. Pressured by the United States, Israel withdrew from Sinai in 1957 but on the condition that the United Nations ensured that the Strait of Tiran remained open to its ships. In 1967, when Egypt caused the UN to withdraw from Sharm el Sheikh, Israel launched a blitzkrieg that overnight gave it control of the whole of Sinai to the Suez Canal. In the so-called Six Day War, Israel also took the West Bank from Jordan and the Golan Heights from Syria. From then on to the Yom Kippur war of 1973, Israel, with access to the Red Sea through the Strait of Tiran, closed the Suez Canal and thus denied Egypt of its principal means of earning foreign currency. Israel turned the Suez Canal, a seaway of global importance, into a moat for the defence of its newly-acquired territory of Sinai. The settler-state planted many mines in the Suez Canal and built the Bar-lev line on the east bank of the seaway.

The resumption of hostilities in the so-called War of Attrition for the Suez Canal, with Egypt shelling Israeli positions on the east bank, and Israel retaliating with its air force, brought Cairo to its knees. With millions of refugees fleeing from the flattened towns of the Canal Zone, and with Israeli planes roaming at will over the cities of the Nile valley and delta, Egypt only avoided another crushing defeat by the timely intervention of the Russians. Soviet military experts erected a sophisticated missile air-defence system in record time. The Russians got little or no thanks for their assistance; no sooner had they completed the deployment of the vital missiles than the Egyptians turned them out. The Russians had become very quickly, through arms supply and military personnel, the dominant external factor in Egyptian politics, and they were keen to entrench themselves alongside a waterway which would be of the utmost importance to their maritime strategy when it reopened. The Egyptian leaders, however, soon felt that the presence of the

Soviets was becoming overwhelming and that they were meddling in their internal politics. Saudi Arabia, that had become Egypt's paymaster since the closure of the Suez Canal, was also pressing Cairo to distance itself from the Russians. With the VLCC tankers taking oil to the West by the Cape route, Suez was no longer economically vital. Militarily also, the West was losing interest in persuading Israel to come to an agreement with Egypt about reopening Suez since this would give a relative advantage to the Soviet Union. The Egyptian leaders felt that it was urgent to take a new initiative.

Egypt's move was to use the substantial military equipment it had received from Moscow for a surprise attack across the Canal, synchronised with an attack by Syria on the Golan Heights. These military moves were combined with a simultaneous diplomatic offensive, using the Arab oil weapon to persuade the West to get Israel to make concessions about the Suez Canal. The Israelis had counterattacked and succeeded in getting some of their forces to the west of the Suez Canal by the time that a UN ceasefire was accepted. In 1974, Israel and Egypt agreed to a disengagement whereby Israel withdrew all of its forces from the west side of the Suez Canal while Egypt retained a limited force on the east side of the seaway. The 1974 disengagement marked the beginning of the end of the Israeli-Egyptian confrontation and led eventually to Camp David. Having saved face by their relatively good performance in the Yom Kippur War, the Egyptians had enlarged their diplomatic offensive by going further than harnessing Arab oil power to influence the West: They offered peace and recognition to Israel in exchange for its complete withdrawal from Egyptian territory. That was a real breakthrough from the point of view of Israel, because without Egypt, by far the largest military power in the Arab world, the balance of forces would be so overwhelmingly in favour of the settler-state that its opponents would no longer pose a credible threat to its existence. The principal gains for Egypt of peace with Israel have been the recovery of its territory of Sinai, albeit without the Gaza Strip, and, above all, the reopening of the Suez Canal. On the ·30 April 1979, the first Israeli ship ever to sail from Eilat to Haifa was given a ceremonial escort by the Egyptian navy. Egypt paid a price for recovering its territory and canal dues. It forfeited the role of the leadership of the Arab world and alienated itself from Arab nationalism. In 1981, President Sadat paid with his life the handshake with Menahin Begin on the White House lawn.

Right from the start, the settler-native conflict was not confined to Palestine. The Arab states and public opinion throughout the Arab world identified with the native Palestinians. An international sub-system developed on that native-settler conflict that polarised on the Cold War, with the United States supporting Israel and the Soviet Union backing the Arabs. A fully-armed Israel was an asset in the military containment of the Soviet Union policy by the West in the Middle East. Politically, however, it was a liability as it facilitated the Soviet Union penetration of the Arab world, leap-frogging over the Baghdad Pact-Cento. Moscow compensated, to some extent, for its major reverse in Egypt in 1973 by strengthening its influence in Syria and in Iraq. Further south in the Red Sea, when Moscow failed to reconcile Somalia, where it had a military facility in Berbera, with the much larger Ethiopia, it chose the latter and helped it to crush the Eritrean secessionists. The Russians also cultivated their relations with the radical South Yemen where they got the use of Aden. South Yemen also owned the island of Perim in the middle of the strategic chokepoint of Bab el Madeb. The strait of Bab el Mandeb was not only crucial for the Suez route but also for the Israeli alternative route via the Gulf of Akaba. With its strong position at the southern end of the Red Sea, the Soviet Union acquired leverage on Egypt to ensure passage through the Suez Canal for its commercial and naval vessels. In 1980, Soviet ships carried 1000 Cuban troops for Ethiopia via the Suez Canal. Admiral Gorchkof's navy went back and forth between the Black Sea and Vladivostok using the Suez Canal.

The international sub-system around the native-settler conflict in the Middle East has not disappeared with the end of the Cold War. Without the presence of the Soviet Union, and with Egypt having made peace with Israel, the balance of forces in the sub-system is overpoweringly in favour of the settler-state. Palestinian radicals have adopted some of the tactics that served the nationalists well in wars of decolonisation. In these asymmetrical confrontations of decolonisation, the nationalists avoided losing militarily for long enough for the colonial state to come to the conclusion that it stood to lose more politically by continuing a war which it could not win once the guerrillas were "like fish in the sea" in the midst of the mobilised population. Systematically eliminating the natives was not politically feasible. Large-scale genocide would have defeated the purpose of the colonial state for the aim was not to retain the land through the physical elimination of the natives, but to win them over to the side of the government against the radical nationalists. In the end, the nationalists triumphed because the cost in political terms of their military elimination was too high for the colonial authorities. When the allegiance of the native inhabitants was at stake, rather than the retention of territory, David defeated Goliath.

The upside-down relationship between military force and political power, which obtained in the rare cases where large-scale violence was present in the decolonisation of non-settler colonies, does not work in settler-states. While it was possible for the imperial power to consider loss of territory as a relative victory if by granting independence it could still influence the native nationalists, the loss of the land in a settler-state would put an end to its existence as a sovereign state, and perhaps as a people. The use of violence by the natives in Palestine, without the support of Arab states, is unlikely to sway the settlers. The asymmetrical violent confrontation between natives and settlers in Palestine, however, is detrimental to the policy of the now sole Superpower in the international system. Without the Cold War polarisation, the United States has more room to manoeuvre and does not have to give unconditional support to the settlers. To keep alive the "peace process" and get more Arab states to follow the Egyptian example, the United States must get the settlers to make concessions to the natives in Palestine. There are limits, however, to the concessions that the settler-state can be pressed to make. The settlers have sufficient "kith and kin" support in the United States to ensure that they will not be pushed to abandon what they consider to be their vital interest. The "two nations in one state" solution: a Jewish nation and an Arab nation in a federal Palestine is totally unacceptable. The constitution of Israel gives Jews the world-over the "right of return" and immigration has swelled the settler's population. Non-Jews cannot be integrated into the state as full citizens. In this respect, the minority of Arab Palestinians who stayed in Israel when it was founded are not an asset but an embarrassment; indeed, they are perceived as a potential threat. The "two sovereign states in one Palestine" solution, which is now acceptable for the moderate natives, is not preferable to the status quo for the settlers. Yet, unless the Arabs are adequately appeased over the core issue of the settler-native conflict, the "peace process" is not irreversible. "Rogue states" in the Arab world and beyond - Iran is a case in point - will want to acquire weapons of mass destruction as long as Israel is armed with nuclear weapons and receives massive military aid from the United States. "Draining the swamps" will be a never-ending toil if a just and acceptable solution is not found for "the native problem" in Palestine.

Summing up

Colonisation was closely linked with predominant power at sea in the Euro-centric international system. Decolonisation, in the global international system, which was polarised on the two Superpowers, the United States and the Soviet Union, never meant that the West was abandoning power at sea. On the contrary, with the Soviet Union as the predominant land power in the Eurasian land mass, sea power became more - not less - important for the West. Nowhere was this truer than in the Indian Ocean. Hegemonic power at sea, and colonies in virtually all the littoral, had turned the Indian Ocean into a British lake. By the time that economic constraints caused Britain to relinquish its last footholds East of Suez, the United States was well entrenched in the Indian Ocean with a pivotal base in Diego Garcia and alliances with the settler-states in the offing of the seaways.

The almost landlocked configuration of the Indian Ocean gives the three passages, the Cape, Malacca, Suez, which link it to the rest of the global ocean, considerable commercial and military significance. Sea power can be vulnerable in these chokepoints. Three European settler-states - South Africa, Australia, and Israel - were implanted in the offing of these all-important seaways. All three settler-states were faced with a "native problem" internally that had to be solved in one form or another to make the settler-state irreversible. Of the three settler-states, Australia is the only one that early on found an irreversible solution to "the native problem". The solution that South Africa adopted, apartheid, failed. The natives of Palestine have not accepted as final the solution imposed by the Zionist settlers. International sub-systems developed around the settler-native conflicts in South Africa and Israel as these conflicts spilled over into neighbouring states and interacted with the east-west conflict in the Cold War. The end of the Cold War coincided with the end of apartheid in South Africa and the Southern Africa sub-system has disappeared. The United States, as the sole Superpower of the present international system, has a determinant role in the transformation of the sub-system around the settlers-natives conflict in the Middle East.

South Africa is on the Cape route, unlike Australia and Israel, which are in the offing of Malacca and Suez but not directly on these seaways. While the West has hegemony at sea there cannot be a threat to the Cape route from the sea. Threats to the Cape route would have to come from the land, from a South Africa in the hands of an enemy of the West. South Africa posed no threat to the Cape route during the apartheid rule of the settlers. South Africa now, with the natives in political control, is a factor of stability in Southern Africa. A failed or failing state in South Africa harbouring trans-national anti-West terrorists that could mount a land threat to the Cape route is unlikely.

Australia collaborates with Malaysia, Singapore, and Indonesia to deter and ward off threats to the Malacca route. There can be no large-scale military threat to that route from the sea so long as the United States is the thalasocratic hegemon. Conventional threats from the land by a great power are non-existent for the time being. In the future, China could pose such threats as part of a confrontation challenging the hegemony of the United States. More likely for the time being, however, are unconventional threats such as terrorism and piracy, as well as ecological threats.

The failure to find an acceptable solution to the settlers-natives conflict in Palestine has had a detrimental effect on the Suez Canal route. Peace with Israel and realignment with the West enabled Egypt to reopen the Suez Canal. With international assistance, Egypt has enlarged the Suez Canal with the aim of attracting back the oil traffic. The Suez Canal route is unlikely to be once again the main oil route to the West. Economies of scale, as well as security and ecological considerations, keep the giant tankers on the Cape route. Even if more Arab states follow Egypt's example of making peace with Israel, the long narrow corridor from Bab el Mandeb to Port Said will remain vulnerable to terrorism for as long as the "peace process" fails to find a solution to the settlers-natives conflict over the land of Palestine.

Chapter 3

The Strategic Values of the Indian Ocean to Indonesian Diplomacy, Law and Politics

Hasjim Djalal

Introduction

The Indian Ocean has so many strategic values for Indonesia, either in history, culture, defence, security, economy, and in law and politics. In the field of law, especially in the law of the sea, Indonesia has thousands of miles of coastline on the west of Sumatra and the south of Java. Except with Australia, including with the Christmas Island and at some points between India (Andaman and Nicobar Islands) and Aceh, Indonesia has open maritime boundaries with the Indian Ocean, either for the purpose of "Straight Archipelagic Baselines" or for Territorial Sea limits as well as for the Contiguous Zone, Exclusive Economic Zone and Continental Shelf. Almost all of these maritime boundaries could in principle be determined unilaterally by Indonesia, except the outer limits of the Indonesian Continental Shelf if the geology and the geomorphology of the seabed area westwards and southwards of Indonesian Islands still permit Indonesia to claim Continental Margin beyond 200 miles from its "Straight Archipelagic Baselines". Thus, the Indian Ocean contains enormous economic, political, and strategic potential for Indonesia that are so important for the future of the nation and the state.

Indonesia and Indian Ocean Fisheries

In the field of fisheries exploitation and management, the Law of the Sea Convention 1982 has stipulated that "straddling fish stocks" as well as "highly migratory species" like tuna necessitate cooperation, either bilaterally between the coastal states and the far-distant fishing states, or through regional cooperation organizations, or arrangements between the coastal states and their neighbours and the distant-water fishing states, in order to be able to manage and utilize those fisheries resources sustainably. In the Indian Ocean Region, there already exist organizations, particularly the Indian Ocean Tuna Commission (IOTC) under the aegis of the FAO, whose headquarters is in Seychelles. Although Indonesia is a full member of the FAO, it is regrettable that Indonesia up to now is still not yet a full member of the IOTC, although it has attended the meetings of the IOTC as a cooperating non-member. In fact, there are distant-water fishing states, such as Japan, China, South Korea, European states and other coastal countries that have long become full members of the organization and have been playing an important role in the management of these fisheries resources. It is to be hoped and to be encouraged that Indonesia will become a full member of the IOTC in the near future in order to be able to protect its own fisheries resources in its own EEZ and, at the same time, promote its interest in the management and sustainable utilization of the fisheries resources of the high seas beyond the EEZ in the Indian Ocean.

Similarly, with the case of the Convention on the Conservation of the Southern Bluefin Tuna (CC-SBT), whose headquarters is in Canberra. While the SBT generally spawn in the Indonesian EEZ south of Java and that the fishing states like Australia, New Zealand, Japan and South Korea (Chinese-Taipei may soon join the CC-SBT) have become members, Indonesia so far is not yet a member. In my view, this could endanger Indonesia's own interests, particularly in the management and conservation of these very highly expensive fish in Japan. It is to be hoped and to be encouraged so that Indonesia could become a member of CC-SBT in order to be able to participate in the management, exploitation, and conservation of those highly valued natural resources in the Indian Ocean, either for the Indian Ocean as a whole or for its own interests.

Indonesian Seabed Jurisdiction

On the seabed area and its subsoil in the Indian Ocean beyond the limits of national jurisdiction, namely, beyond the outer limits of the Continental Shelf in the continental margin, according to the Law of the Sea Convention, these resources would be managed by the International Seabed Authority (ISA) in Jamaica. Up to this moment, Indonesia has not yet determined or claimed any of its continental shelves beyond 200 nautical miles from its straight archipelagic baselines, although Indonesia has begun a study to examine these possibilities. So, for the moment, the outer limit of the Indonesian Continental Margin in the Indian Ocean is still not yet determined. According to the Law of the Sea Convention, Indonesia has the potential, after conducting the necessary research and survey, to submit claims for the continental margin if it still has the margin beyond 200 miles following the stipulations and conditions set out in the Law of the Sea Convention. Again, according to the Law of the Sea Convention, Indonesia could submit a claim within 10 years after the entering into force of the Convention for Indonesia, namely before November 16, 2004. I understand, though, that the time limit has been extended for 5 years, namely until November 16, 2009. The claim is to be submitted to the Continental Shelf Commission (CSC) in New York. In fact, the CSC could help coastal developing countries to study their possibility for claims to the continental margin. So far, Indonesia has not yet requested the assistance of the UN in this respect. I personally hope that Indonesia would pay more attention to this possible claim and submit the claim before 2009, after careful studies, if there is still any continental margin of Indonesia beyond its 200 miles EEZ.

As we all know, Article 76 of the UN Convention on the Law of the Sea 1982, allowed for the determination of the outer limit of the Continental Shelf beyond the Territorial Sea limits of the "natural prolongation" of the land territory of the coastal state to the Seabed Area, using either one of the following criteria:

- To the distance of 200 nautical miles from the baselines from which the breadth of the Territorial Sea is measured, which in the case of Archipelagic State like Indonesia, from straight archipelagic baselines.

- To a fixed point at each of which the thickness of sedimentary rocks is at least 1% of the shortest distance from such points to the foot of the continental slope.

- To the distance of not more than 60 nautical miles from the foot of the continental slope. The distance indicated in the second and third above shall not exceed 350 nautical miles from the baselines, or not more than 100 nautical miles from the 2500 m isobath.

Thus, it is really necessary to conduct serious research and surveys on:

- the location of the foot of the continental slope of Indonesia in the Indian Ocean

- the general outline of the 2500 m isobath

- the maps regarding the thickness of sedimentation and its distance from the foot of the continental slope.

In the Indonesian EEZ, including in the Indian Ocean, Indonesian jurisdiction is not limited only to fisheries, although fisheries is one of the most important components of the EEZ. The UN Convention on the Law of the Sea 1982, in Article 56 stipulated that in the EEZ Indonesia has "sovereign rights" on the natural resources, including fisheries and energy from the sea, and jurisdiction with regard to the protection and preservation of the marine environment, marine scientific research, and the establishment and use of artificial islands, installations and structures. There are a lot of things which need to be done by Indonesia and by other coastal states to make use of all of these possibilities.

In the EEZ and the air space above it, there is freedom of navigation and overflight. Nevertheless, up to now, there is no clear legal certainty on the use of the EEZ and its airspace for military purposes, either for military exercises or for intelligence gathering activities and espionage, either by using radar technology, satellites, or military flights. Although regulations on non-military navigation and civil aviation have been made separately, either through the IMO and ICAO, there are still a lot of things that need to be studied and agreed upon on the military activities of military vessels and aircraft in the EEZ of other countries. These problems have now become major topics of discussion in various circles, particularly because of different interests involved among some major coastal countries and major maritime powers, including in the Indian Ocean. In view of the increasing significance of the Indian Ocean, particularly because of the volatility of the situation in the oil-rich Middle Eastern countries, Indonesia and other coastal countries in the Indian Ocean should pay more attention to this matter in order to protect their political and security interests in the Indian Ocean, including in their EEZ.

As indicated above, in accordance with Article 33 of UNCLOS 1982, Indonesia could also determine its own Contiguous Zone in the Indian Ocean up to 12 miles beyond its Territorial Seas for the purposes of controlling violations of its customs, fiscal, immigration, and sanitary laws. Officially, Indonesia has not yet declared its Contiguous Zone in the Indian Ocean, although many people, by implication, assumed that by ratifying the UNCLOS 1982, Indonesia in principle has adopted the Contiguous Zone principles. I would particularly hope that the Indonesian Government would pay more attention to the establishment and utilization of the Contiguous Zone concept in order to prevent smuggling of goods and arms, illegal trafficking of drugs and persons, maritime terrorism, and other crimes that are recently rampant in some parts of the Indian Ocean Region.

Beyond the EEZ, there are the High Seas where freedom of navigation and overflight of all states are recognized, including for Indonesians. Those freedoms include freedom of fishing, freedom to carry out marine scientific research, to lay down cables and pipelines in the seabed areas, or to construct artificial islands and other installations at sea. However, recently, there are many rules in International Conventions that regulate those freedoms, either in the field of navigation, fisheries, marine scientific research, or even protection of the marine environment. Currently, many new concepts are being developed and discussed in various academic and government circles, specifically regarding management of the ocean based on ecosystem and Large Marine Ecosystem principles that may transcend the maritime boundaries of a state. Indonesia and other coastal countries of the Indian Ocean need to be more proactive in these activities in order to protect their own interests as well as the sustainability and the proper management of Indian Ocean resources and their ecosystems. I notice that a number of countries, such as India, Australia, Japan, China, and South Korea have played an active role and have participated in various marine scientific activities globally, including in the Indian Ocean.

In the International Seabed Area beyond the Continental Shelf in the Indian Ocean, the management of its natural resources, particularly minerals, is regulated by the

International Seabed Authority in Jamaica. Up to now, only India has obtained exploration rights in the Central Indian Ocean basin in order to look for and exploit the nodules in that area, which may be rich in nickel, copper, cobalt, manganese and other minerals. In addition, new types of minerals on and in the seabed have also been discovered, particularly metallic sulphide that have been found in the volcanic parts of the seabed area (geothermal vents) and metallic crust, which are discovered on the surface of the sea mounts in the seabed area. Both sulphide and crust contain similar minerals to the nodules. The International Seabed Authority in Jamaica has prepared regulations on the prospecting and exploration for these new minerals. It should be noted, however, that the exploration of the seabed area could affect the maritime environment, particularly the life of the biota on the seafloor and in the bottom water column. In addition, other resources, such as methane hydrate have also been discovered, particularly in the subduction zone. At the same time, new living resources have also been gradually discovered (and exploited) on the seabed area and in the deep water column. Indonesia, unlike India, China, South Korea and Japan, has not yet given serious attention to the potentials of these new minerals and other resources from the seabed area. Indonesia is still concerned with how to protect the exploitability of its land resources from the possible competition from these deep seabed mineral resources. In any case, Indonesia will continue to play an active role together with other states around the Indian Ocean in the operation of the International Seabed Authority in Jamaica.

The Indian Ocean as a Zone of Conflict

From a political point of view, the maritime jargon so far has been that the Atlantic Ocean is regarded as the Ocean of the past, the Pacific Ocean as the Ocean of the present, and the Indian Ocean as the Ocean of the future. Now, the Indian Ocean has approached its future and may have become already the ocean of the present, especially after the Soviet Union and its communism is no longer a major power in world politics and in the military equation. Thus, especially after China also practices market economy, many observers begin to see the emphasis of global politics and conflicts spreading from the Pacific Ocean to the Indian Ocean. Several factors may have caused this change, especially the oil factor as a global strategic material, either for the economic and industrial growth of East Asian Countries or for military purposes.

Within the context of this oil politics, the Middle East remains one of the most important sources of energy. In addition, Central Asia, such as Kazakhstan, and Central Siberia, are beginning to become new sources of energy. Thus, the new struggle would be how to protect the oil resources in the Middle East and Central

Asia, either for the interests of the United States and its allies or for China and Western Europe, as well as for India. One of the most important routes for oil from Central Asia is through the Indian Ocean, and even this problem has created some kind of competition between Iran and Pakistan as well as Turkey.

It should also be noted that the Middle East is also the cradle of Islamic culture that has influenced the development of the states around the Indian Ocean, either in East Africa, South Asia, or South East Asia. The Moslem population in these areas, particularly in the Middle East, is increasingly regarded as having the potential to upset the supplies of the oil to the world, particularly with the increasing activities of international terrorism that many have accused some radical Islamic elements living in the area around the Indian Ocean as some of the instigators. Thus, the strategic significance of the Indian Ocean in the field of politics and security is increasing, although economically none, except Australia, of those countries belong to the advanced industrial countries. India is just beginning to be a major global economic power and it is expected that it will play a major and constructive role in the Indian Ocean. Generally, however, economic growth and economic conditions of the states around the Indian Ocean are far below the growth of countries in other parts of the world, especially East Asia, Europe, and North America. Some analysts say that the situation of economic backwardness could incite jealousy towards the West and advanced countries that in the end may bring radicalism and perhaps feed on international terrorism. This situation could worsen if the Western world, particularly the United States, is perceived to be fighting against the Islamic world and do little to promote the economic development and well-being of the developing countries.

Indonesia and Indian Ocean Regional Cooperation

Over the years, Indonesia has actively participated in the development of regional cooperation in the Indian Ocean. In fact, the first Asian-African Conference in Bandung in 1955 was motivated by the desire to bring independence and economic and social progress to the countries of Asia and Africa in which the Indian Ocean was basically regarded as "the bridge" for the two continents. Later, with the advance of maritime policies and the inception of the Law of the Sea Conference, the role of the ocean for economic progress and development is increasingly felt in the Indian Ocean Region. Several efforts had been made and the Indian Ocean Marine Affairs Cooperation (IOMAC) whose Charter was adopted in Arusha, Tanzania, in 1991, was activated with its headquarters in Colombo. This organization was very active for some time in promoting marine affairs cooperation in the Indian Ocean, but later on seemed to be "losing steam" in its activities. Later on, another organization was established, namely, the Indian Ocean

Rim Association for Regional Cooperation (IOR-ARC) with its headquarters in Mauritius. This organization, concentrating on cooperation among Government officials through Senior Official Meetings (IOR-SOM), on cooperation in trade and investment through the National Chamber of Commerce and Business Forum (IOR-BF), and on scientific and academic cooperation through the Indian Ocean Rim Academic Group (IOR-AG). It seems that the IOR-ARC has achieved some progress, but there is still a lot that needs to be done to realize and implement the agreed modalities. In the meantime, various academic groups have also been active in the Indian Ocean, including the Indian Ocean Centre in Perth, Australia (which now is no longer active) and the current Indian Ocean Research Group (IORG) with its "twin pillars" in Chandigarh, India and Perth, Australia. Other organizations, such as the Indian Ocean Tourism Organization (IOTO) have also attempted to promote tourism among Indian Ocean countries. Even the UN Indian Ocean Committee at this moment no longer seems to be very active. It is therefore very important to examine and highlight the reasons why these efforts in the past have "come and gone".

Indonesia's Strategic Situation in the Indian Ocean Region

Indonesia has one of the most strategic positions with regard to the political and maritime development of the Indian Ocean, not only because it has a very long coastline in the Indian Ocean, but also because its very strategic geographical position between the Indian and Pacific Oceans, through which the Western countries, especially the United States, Australia, Japan and South Korea require safe and secure passage through Indonesian maritime zones and air space. In the future, the entry points to the Indian Ocean and the Pacific Ocean through Indonesian waters in the Straits of Malacca and Singapore, the archipelagic waters in the Karimata and Sunda Strait, as well as through the Straits of Makasar and the Straits of Lombok, and the Indonesian Eastern Seas (Moluccas, Seram, Banda, Arafura, and Sawu), will become more significant. It is expected that the increasing role of these strategic sea routes in the future will give Indonesia a much more significant political and security interest in Indian Ocean political, economic, and social development.

Thus, the Indonesian strategic position is actually very strong in the Indian Ocean region for at least seven principal reasons. First, because of its very strategic position between two oceans and two continents, namely, between Indian and the Pacific Oceans, and between the Australian and Asian continents. Second, the geographical structure of Indonesia that largely consists of seas, which are necessary for the global strategy of the maritime powers who have interests in the Indian Ocean (at the same time this geographical condition is also making

Indonesia "vulnerable" to foreign intervention). Third, Indonesia is a very rich country with strategic resources, which are also important for the world industrial powers. Fourth, the very close historical and cultural relations between Indonesia and the cultures of the Indian Ocean: Hinduism, Buddhism, Islam, as well as Christianity. Fifth, Indonesia is now the third largest Democracy in the world, the fourth most populous state in the world, among the ten largest countries in the world, and its Moslem population is the largest in the world while the Moslem communities are among the largest moderate Islamic groups in the world. These factors should contribute toward positive roles that Indonesia will be able to play in the development of political, economic, and socio-cultural progress in the Indian Ocean Region. Sixth, Indonesia is among the countries that have obtained independence through armed-struggle after the Second World War, one of the initiators of the Asian-African Conference and the birthplace of the Bandung spirit which has contributed so much for the independence of states in Asia and Africa. Seventh, Indonesia was also one of the founding fathers of the Non-Alignment policy in the world, a country whose role in the group of developing countries, in the development of regional cooperation, and within the organizations of Islamic Conference, is generally regarded as positive. All of these factors constitute important assets in developing positive Indonesian interests in the Indian Ocean Region.

Conclusion

It is, therefore, very important to understand and to study how best to develop Indonesian potentials in the development of political, economic, trade, culture, as well as legal issues between Indonesia and the Indian Ocean countries for the sake of the future of Indonesia and Indian Ocean states. It is important to study the development of those interests in the context and situation of many Indian Ocean countries, which are suffering economic difficulties in the face of the increasing processes of globalization in all fields; in economics, trade and industry, in the field of politics and democratization, protection of human rights, protection of the environment, as well as globalization, telecommunication and communication. It is important to ensure that Indonesia's promotion of its national interests go hand-in-hand with the interests of other countries, particularly around the Indian Ocean, so that the pursuit of those national interests do not create confrontation with the countries around the Indian Ocean Region. In this context, many efforts to promote regional cooperation have been attempted in and around the Indian Ocean. Yet, it is essential to continue and intensify diplomatic efforts so that all of these efforts become more effective and sustainable in the future.

Chapter 4

The Emerging Centrality of SLOCs in China's Maritime Strategy

Swaran Singh

Introduction

China's ever-increasing dependence on imports such as oil, natural gas, food and technologies have come to be an intractable problem - making China vulnerable to the ever-increasing leverage of external powers that could disrupt the free flow of these goods and services to and from China. Indeed, China's leaders strongly believe that their trillion-dollar annual foreign trade - known as the locomotive of their rapid development - could be held hostage by powers that can flex their military muscles across the northern littoral and sea lanes of the Indian Ocean that provide a critical link between Europe, Africa, the Middle East and the Chinese ports across the South China Sea. Moreover, the rising international maritime traffic, maritime terrorism and the expanding similar interdependence of other regional players such as India, Japan and Korea puts China's requirements for ensuring freedom of navigation in the Indian Ocean sea lanes in direct competition with all other regional powers and thus further compounds Beijing's policy options.

China is also known to have maritime disputes with all of its maritime neighbours. This continues to mar both the evolution and implementation of China's maritime vision, so integral to its peaceful development in the coming years. Even if Beijing has since shed its old rhetoric of sovereign rights and military means and come to focus on joint economic development of the South China Sea, China has not completely given up on its occasional muscle flexing although articulated in a far more subtle manner. If anything, China currently follows a double-edged policy of trying both carrot and stick to achieve optimal results (Radio Free Asia, 2005; Taiwan Affairs Office of the State Council, 2005). Even when China may have developed some control and cooperation in the South China Sea, it is the Indian Ocean sea lanes of communication (SLOCs) - connecting China to Europe, Africa and the Middle East - that continue to pose a serious challenge to Beijing's policy planners. Accordingly, it is these SLOCs of the Indian Ocean rim that have witnessed China's increasing indulgence and initiatives, all aimed at developing some noble means of building its maritime power projection capabilities to protect its shipping and other maritime interests (Cole, 2005; Gertz, 2005; Lelyveld, 1997).

It is in this rapidly evolving new context of China's changing vision and equations that this chapter aims to highlight the emerging centrality of these SLOCs in China's overall maritime vision, especially its increasing preoccupation with SLOCs of the northern littoral of Indian Ocean. It is against this backdrop that we will examine China's alternative strategies in dealing with its challenges and its future policies and plans towards ensuring freedom of navigation and a degree of control over these SLOCs which have come to be so critical for the stability and peaceful rise of China (Dong Guozheng, 2004; Zhang Guihong, 2005). The chapter especially explores the possibilities of the Kra Canal project (in Thailand) that not only promises to provide an alternative to the Malacca Straits but would also provide opportunities for China's Yunnan Province and the Mekong basin to emerge as a booming transshipment region with considerable geostrategic implications.

Freedom of Navigation: The New Context

Starting from the early 1980s, China's rising awareness about its maritime rights (for example, exclusive economic zone, territorial seas and freedom of navigation as provided under the Conventions on the Law of the Sea) was to change China's maritime thinking altogether (Cooper, 2000; Valencia, 1997, p. 263). This was to be reflected in the fundamental shift in China's inward-looking naval doctrine of sea guerrillas to its strategy of 'offshore defence' (yinyang fangyu) which was aimed at not just defending China's coasts but at preventing incursions into China's coastal waters (Tai Ming Cheung, 1990; You Ji and You Xu, 1991). This was to result in China redefining its 'strategic frontiers' and laying claims to the whole of the South China Sea and beyond (Shambaugh, 1994; You Ji, 1995). This was also to see China exploring alternatives to its rising imports of food grains and energy (for example, fish and oil and gas from its seas), looking for alternate suppliers (like Russia and Central Asia) and evolving alternate supply routes (such as energy pipelines).

From the early 1980s, with the rise of China as the next global power on-the-horizon, China's expanding power profile was also to add to its responsibilities. For instance, ensuring freedom of navigation not only for its merchant vessels but also for its warships aimed at effecting freedom of navigation and protecting critical SLOCs is seen today as the lifeline of China's security and development. However, China's success in exploring these alternative channels to sustain equilibrium between its ever-increasing surplus and deficit in both goods and services has been limited at best. As a result, China's increasing dependence on its traditional land-based supply lines or SLOCs continues to rise unabated with no alternative channels in sight. Similarly, China's continued dependence on its traditional suppliers - mostly from Europe and the Middle East - further reinforces the significance of Indian Ocean SLOCs, which are seen today as critical for sustaining China's juggernaut of rapid economic development and which remains the pre-requisite for China's political stability and social cohesion. Thus, awareness about its rights and needs over ensuring freedom of navigation marks a new context that makes SLOCs so central to China's 21st century maritime vision.

Conversely, this period has also witnessed China's naval prowess becoming noticeable from the early 1980s. There have thus been consequent debates over China-threat theories explaining how China could potentially use its maritime power to disrupt the SLOCs for its not-so-friendly neighbours in order to seek compliance with its wishes. This period, for example, has witnessed the revival of China's claims to the whole of the South China Sea and the Chinese navy becoming far more active occupying islands, reefs and islets throughout this disputed water body of the South China Sea. This has also witnessed China flexing its military muscles against Taipei (Campbell and Mitchell, 2001). Accordingly, it is believed that in case of any military stand-off across the Taiwan Straits, China could take drastic measures blocking SLOCs to arm-twist its adversaries. So much so that Beijing had to issue clarifications repeatedly saying that:

... while safeguarding its sovereignty over the Nansha [Spratly] islands and its maritime rights and interest, China will fulfill its duty of guaranteeing freedom of navigation for foreign ships and air routes through and over the international passage of the South China Sea according to international law (Valencia, 1997, pp. 274).

In view of China's increasing economic interests and rising stakes in SLOCs and also its gradual engagement with its neighbours, not many believe that China would be seriously contemplating such a disruption to be its preferred policy. Besides, the situation remains far more complex when it comes to navigational rights and obligations of the military warships (Roscini, 2002). In this context, it is the United States that has a global presence and stakes in all SLOCs, and one that must ensure freedom of navigation for its warships and submarines through and under the SLOCs of the Indian Ocean. Looking from a US perspective, four of the 16 strategic straits around the world are in Southeast Asia - Malacca, Lombok, Sunda and Ombai-Wetar - where China can potentially pose a major challenge and use SLOCs for arm-twisting. As a result, the US has repeatedly expressed its commitment to uphold and defend the freedom of navigation and maintain strategic SLOCs linking Northeast Asia to Southeast Asian and Indian Ocean SLOCs (US Department of Defense, 1995).

Even from a purely theoretical perspective, while open naval blockade seems less likely, several experts believe that 'mining' of critical SLOCs may emerge as a stronger possibility in near future. However, even this has been seriously questioned over time. While sea-mining can be a credible threat in coastal waters and the relatively shallow Malacca Straits, the depth and currents of Sunda and Lombok Straits would radically reduce the effectiveness of sea-mines. Besides, the technologies in detecting and clearing mines have also advanced though possibilities of a terrorist strike have increased. However, the possibilities of crisis in the Taiwan Straits leading to military disruption of commercial vessels in this region remain another possibility. While the question of the impact of such disruptions on SLOCs has remained only academic, such sporadic hype has repeatedly had a real impact on insurance and diversions pushing several countries including China into constantly exploring for alternatives. And here, looking into China's history may have some indicators for future possibilities.

China's SLOC Consciousness: Genesis and Evolution

Very briefly, China has one of the oldest maritime traditions in the world. Some scholars have traced this as far back as the Warring States period in 221BC, but its maritime consciousness has remained sporadic and fluctuating (Fairbank, 1983, pp. 24-27; Reynolds, 1989, pp. 3-11). Especially during the 12th to 15th centuries (under the *Song*, *Yuan* and early *Ming* empires) China is known to have had a glorious naval history that clearly underlines its SLOCs consciousness, though it did not emerge as a major seafaring state in post-Industrial Revolution modern times (Chan Hol-lam, 1971). The high point of China's naval expansion had occurred during the Song dynasty period (960-1279 AD) when its first 'national' navy was established which was then the most powerful and technologically most advanced in the world (Gang Deng, 1999, pp. 169-170). With more than 10,000 vessels and 50,000 sailors, China was to remain the sea power until the early 17th century or so, leading the world in maritime commerce, shipbuilding, voyage management, navigational sciences and oceanography. Large fleets were recorded to having launched for invasion of foreign territories like Viet Nam, Java, Japan and so on thus expanding China's maritime commerce. A 15th century, epical hero, Muslim Admiral Zheng, was to lead expeditions as far as the Middle East and Africa and anchor at several ports in South and Southeast Asia.

From then on, it is the European Industrial powers that were to overtake seafaring. As a result, China's navy was to play a disastrous role in the Sino-Japanese war of 1894-95 and also have no role in China's campaigns during World War II. It was only after a long break - since the Portuguese and other European powers moved their navies to steel ships with steam engines and launched a new era of colonial expansion setting in the decline of China - that Beijing was to once again revive its maritime vision from the early 1980s (Singh, 2003, pp. 105). And since then, given its economic rise, this maritime vision has come to be focused primarily on ensuring the access and safety of SLOCs. Also, though national sovereignty and territorial disputes remain critical components of its naval power-driven maritime vision, yet this has been increasingly expanded to pay greater attention to other larger dynamics of sea borne commerce and interdependence thereby making SLOCs as the new high priority area for China's maritime planners. Besides, in this new context of the post-9/11 period, the increasing threat from piracy and maritime terrorism is providing a further boost to China's activism in this arena (Roy, 2001, pp. 60-61).

As of now, in view of the increasing occurrence of piracy in and around the Malacca Straits during 2002-2004 and the follow-up great power response, the new thesis of the Malacca Straits Dilemma (MSD) has come to be the new focus of Chinese debates on SLOCs. As outlined by President Hu Jintao himself, it is premised on the belief that "certain major powers" were bent on controlling the Malacca Straits to China's disadvantage (Storey, 2006, pp. 1). Accordingly, Beijing has devised a three-pronged strategy to deal with its Malacca Strait Dilemma, that includes: (a) reducing import dependence through energy efficiencies and harnessing alternative sources; (b) investment in the construction of pipelines that bypass the Malacca Strait; and, (c) building credible naval forces capable of ensuring the security of China's SLOCs (Storey, 2006, pp. 3). However, in the face of resource limitations, China's navy finds it difficult even to maintain a simple regular presence on the high seas, let alone maintaining any 'superiority' to ensure control and influence along such widely-flung SLOCs. Meanwhile, China has adopted an alternative strategy of building a series of naval engagements which is driven by this emerging centrality of SLOCs in China's maritime vision. Accordingly, knowing full well that the People's Liberation Army Navy (PLAN) will not be able to rely on its naval prowess alone to protect its vital SLOCs, China has evolved new methodologies to engage a whole range of littoral states in diplomatic and economic measures to ensure a steady supply of energy and other critical resources.

Power Projection Capabilities

To briefly surmise China's naval power projection - the ambitious three-tier vision of Admiral Liu Huaqing - that was outlined in the late 1980s with the objective of building a blue-water navy with global reach by 2049 - has remained too ambitious for Beijing (Kondapalli, 2001, pp. 10; Sakhuja, 2004, pp. 57; Beier, 2005, pp. 291; Howarth, 2006, pp. 44-5). As a result, China has achieved only a limited sea control and sea denial (read deterrence) against major maritime powers like the United States. China, however, still continues to be formidable when it comes to its smaller neighbours. With regard to numbers, for instance, PLAN remains one of the largest navies in the world with an estimated personnel of about 265,000 officers and men. This includes a 25,000-strong naval air force, 5,000 marines and a 28,000-strong coastal defence force. In terms of its equipment; it possesses in excess of 1,700 warships, about 69 submarines and over 700 aircraft. However, the bulk of this equipment remains dated thus limiting China's power projections to its coastal waters (Military Balance, 2005; Sharpe, 2005, pp. 765-767; SIPRI, 2005). Modernisation has been slow and compared to the warships in service in Japan or the US or even most Southeast Asian states, it is only the huge numbers that bring China advantage, if at all. If anything, there has been far more improvement in China's maritime doctrines, though the actual capacity in terms of force structures and weapons systems to implement these continues to lag behind (You Ji, 2002, pp. 26-28).

To briefly look into the various components, China's major surface vessels include about 18 destroyers and 37 frigates. Japan's Maritime Self-Defence Force, by comparison, has 41 destroyers and 20 frigates. Much of PLAN's modernisation projects have been slow and not much improvement has been seen in recent years. PLAN's ocean-going capabilities, therefore, remain ill-equipped for a high-threat environment. China's approximately 69 submarines also remain dated and China has only one Xia-class nuclear powered ballistic missile submarine (SSBN) from the early 1980s. Even this Xia-Class is believed to suffer from 'high-radiation, big-noise and easy detectability' and has not been deployed on the high seas for some time (Beier, 2005, pp. 297). Being unable to achieve continuous patrolling projects a weak triad in China's nuclear deterrence. Though some Ming and Song class submarines have been added since 1987, these continue to suffer from limited endurance and high noise levels which makes them vulnerable to sonar detection thus leaving them largely operable mainly in coastal waters. The most advanced combatant in the PLAN inventory has been the Sovermeny-class destroyer from Russia which is armed with eight Raduga supersonic anti-ship cruise missiles and 170 km range Gadfly missiles as well as its Kamov helicopter that enhances its anti-submarine capabilities (Saunders, 2004, pp. 122). In addition, about eight to ten Kilo-class submarines purchased from Russia since 1994 provide a substantive advantage in sea-denial.

China's naval aviation also remains the world's largest, yet, except for its Russian-made Sukhoi-27s and its J-10s and refueling capabilities, the bulk of its fleet remains dated based on Soviet designs of the 1950s. This again limits their capacity to provide air cover not much beyond the South China Sea and even there it can not be done on a sustained basis or against a capable adversary. China's debates and attempts at building or procuring an aircraft carrier - the ultimate in naval power projection - have been ongoing since the early 1990s, yet this does not seem to be happening in the near future. Even in the case of PLAN's amphibious operations - which remain the most likely scenario in ensuring control over the SLOCs in the South China Sea region - China's assault capabilities may have undergone radical re-organization, yet its lift capacity remains limited to South China Sea waters. China has also neglected the Fleet logistics - a key element for geographically dispersed power projection. It is only recently that China has built Fuqing-class and Dayun-class AOR/AOTs.

Institutions as Instruments of Power

It is against this backdrop of the poor quality of many Chinese military units, that the PLA has tried to develop some superior forces for specific purposes. For instance, the elite forces, the 15th Airborne Army and the marines (naval infantry), are known to be very capable and have been used for power projection operations. From the early 1990s, China has had the lift capability to conduct operations with at least two divisions (25,000 troops) well away from its territorial waters. Upon the seizure of a good port, China could then follow it up with a full Group Army transported by the merchant fleet. However, its amphibious operations do not have air support beyond 500 km, though re-fueling has been demonstrated to be possible. The same also remains equally true of its naval support where much of its equipment remains dated and good for coastal operations only (Kristof, 1994, pp. 382; Wortzel, 1994, pp. 171-172; Ross, 1997, pp. 36). While most of its modernisation projects remain slow and uncertain, China is not expected to either build or buy a blue water navy in the near future (Yung, 1996, pp. 32-48). All of this leaves China's naval power with severe limitations to ensure its control even in the South China Sea and on SLOCs beyond this immediate water body.

As a result, China has been working on alternative channels of seeking access and influence to ensure freedom of navigation. To cite a few examples of this new thinking about moving beyond the conventional naval and power projections approach, China has been focused on building national and multilateral institutions as another noble instrument to strengthen its credentials and to emerge as a leader in Asia-Pacific maritime affairs by 2010. For this, China has been trying to emerge as a country in the Asia-Pacific with the most maritime judicial bodies and the strongest adjudicatory capacity to deal with maritime disputes. Starting from its setting up of its first maritime court in 1984 - in view of its booming maritime transport and trade generated by the opening up policy - China today has 10 maritime courts along its coast and the Yangtze River, which form one of the country's busiest water courses. Also, China today talks of its 'blue economy' and about building 'railroads at sea', which indicates the new tenor of China's maritime institution building at home. All of this remains clearly geared towards projecting China as a new kind of a maritime leader with great potential yet no threat to any other state, especially its neighbours.

China also today seeks to underline the essential distinction between commercial and military shipping when it come to its policies towards SLOCs. Logically, given its increasing dependence on SLOCs and also in view of its increasing mutual interdependence, neither the littoral states nor any extra-regional powers should see any benefit from interfering with the movement of commercial vessels along these SLOCs. However, national responses to such ideas are not guided by economic logic alone. In particular, emotive and radical responses become more likely in the context of a military crisis. But more than that, several countries such as Japan and Korea have already designated 'military zones' across their territorial and adjacent waters for foreign military vessels with which these passing warships must comply. Given its growing dependence on energy imports through the sea, Japan even contemplates the safeguarding of its SLOCs up to 1,000 miles from its coast, which itself could become the cause of a naval stand-off by accident if not by design. Besides, archipelagic states like Indonesia also have special rights to designate passage over its extensive SLOCs that cross through its archipelagic waters. All of this places a premium on legal norms and institution-building, which is where the Chinese have been concentrating for the past few years. The other innovative approach has been their search for alternative SLOCs of which debates on the Kra canal project have lately become rather proactive.

Seeking Alternatives: Kra Canal Mega Project

The increasing traffic in the Malacca Straits and recent rising fears about piracy and international maritime terrorism (especially around Indonesia's turbulent Aceh region) has made China (and others like Japan and the US) seriously proactive in exploring alternative sea lanes. Amongst the possibilities for this region, nothing promises to create a bigger breakthrough than Thailand's impending project of digging the Kra canal to link the Indian Ocean to the Pacific Ocean and South China Sea without having to go through the Malacca Straits. As of now, however, this alternative remains a remote possibility. But once realised - like its sister projects of the Panama and Suez canals - this also has the potential to make history and change the fortunes of many countries in its periphery.

To begin with, the idea of the Kra Canal was first conceived in circa 1677, when Thai rulers had invited French experts for feasibility studies. This involves digging a sea-level Kra Canal through peninsular southern Thailand to join the Gulf of Thailand and the Andaman Seas and to bypass the 621-mile narrow Malacca Strait and provide an alternative route connecting the South China Sea to the Indian Ocean. However, after over three centuries, this seemingly last mega-canal project on planet earth appears to have several interested parties and is estimated to cost about $23 billion. But in turn, it is expected to provide a more than 1000 km short-cut between the Indian and Pacific oceans, thereby saving millions of dollars on shipments, and thus countries such as Indonesia, Malaysia, China and the United States have already been negotiating with Thailand for participation (Ronan, 1936; The Age, 2005). However, this 120-km canal project - which is expected to handle over two billion tons of cargo per annum - has also since become a major concern amongst environmentalists adding on further costs of resettlement and ecological rehabilitation.

Beijing has already developed stronger and closer ties with Thailand and sees this as a project of critical importance to its own economy. One can cite several factors that demonstrate how the Kra mega-project has special attraction for Beijing:

- *Firstly*, this can provide an ideal destination for millions of Chinese workers who will likely be unemployed on completion of the massive Three Gorges dam project by about 2009 as also the Beijing Olympics of 2008. Given this surplus building capacity, China is likely to outbid any other country.

- *Secondly*, this will also provide China (and others) with a dependable alternative route to the Indian Ocean rim. Right from its inception in 1949, China has seriously tried several routes to Indian Ocean waters. These include those through Pakistan, Bangladesh, Myanmar (and now) India, but none of these have been cost-effective.

- *Thirdly*, the Kra Canal promises to provide China a certain control over these SLOCs and also strengthen its engagement with Thailand and other ASEAN members (Foot, 1998, p. 426). Thailand, on the other hand, also needs to achieve prosperity in its rebellious southern regions in order to prevent them falling into the hands of Islamic insurgents.

- And *finally*, the Kra Canal has the potential to catapult Thailand into becoming a regional centre for commodities currently trading, offshore financing, transshipment and other support services. Besides, there would be revenues from navigation fees, export tariffs, tourism, land-bridge toll fees, income tax, land development, aviation and shipyard activities.

However, there also remain several problems which must not be underestimated. The Euro-tunnel, connecting England to the European continent, is the only large strategic transportation project put into operation in the last two decades. Numerous long-planned projects of similar scope and strategic significance - like a bridge across the Gibraltar Strait connecting Africa and Europe, a second Panama Canal accommodating larger vessels than the 60,000 dwt class that can transit the present waterway, a dam across the Bering Strait connecting Siberia and Alaska, a high-speed rail link in China from West to East and from there to Asia, and the Kra Canal - remain on hold. On hold as well, or progressing at a snail's pace, are the energy development projects - whether thermonuclear fusion or solar reactor - that could provide a long-term alternative to fossil fuels and energy transportation through SLOCs which promise to become increasingly problematic (Asia Times, 2000).

The other problem with such mega-projects is that their cost runs into tens of billions of dollars and shows no fast return on investment. Indeed, they have very low financial internal rates of return. However, while financing in an age of privatization of infrastructure has become anathema, and the notion that infrastructure projects must pay for themselves has become dogma, infrastructure development continues to languish. This thinking remains shortsighted and is likely to have serious negative economic consequences, especially for countries and regions now lacking in transportation links and energy resources to support their development. But as the world's greatest builder of huge water works, China seems well qualified to embark on a mammoth project such as the Kra canal.

The Three Gorges dam is now operational and has begun to release several thousands of workers. China has also been building the world's longest bridge connecting Shanghai over miles of water with rapidly developing Ningbo. China has also carried out construction projects in Eastern Africa, and enjoys the requisite goodwill which promises to obtain it a substantial role in the Kra canal project as and when it becomes a reality. This will certainly also obtain China a significant influence and say in ensuring freedom of navigation in these SLOCs and in ensuring free access and safety of these sea lanes to sustain its development and peace. And as China marches towards these promises, its noble and alternative methodologies will play a critical role in giving Beijing an edge over other interested players around its periphery.

Conclusion

To conclude, therefore, apart from historical and geographical dimensions, China's rising consciousness about its stake in the Indian Ocean SLOCs is understood to flow from its: (a) ever expanding dependence on critical imports of energy sources and food grains; (b) its one-trillion plus dollars and still rising (mostly seaborne) foreign trade in general; and, (c) its rising power profile which precludes its expanding engagement with the entire region and other bigger powers operating in its periphery. Though China's indulgence with Indian Ocean SLOCs is viewed by many with concern, yet this remains rather subtle and driven mainly through a three-pronged alternative strategy of: (a) building economic engagement with littoral states along these SLOCs from the Middle East to the South China Sea; (b) getting involved in local infrastructural projects (especially defence-related projects like port dredging and so on) of these littoral countries; and, (c) engaging the local power elite and deploying Chinese military personnel as technicians or advisors in the building of naval facilities for China in the long run. The recent Pentagon report describes this as China's 'String of Pearls' strategy (Hamilton, 2005).

Meanwhile, China continues to face its share of difficulties. For example, the best approach to ensuring security of these SLOCs obviously lies in multilateral cooperation among interested states, yet the issue of SLOCs continues to arouse different response strategies among different countries. Even among various constituencies that have stakes in ensuring the freedom of navigation, the driving motivations and logic remain widely varied. To a military analyst, for example, the SLOCs remain the critical maritime instruments of power, and maritime geography the pivot on which force build-up and deployment must be planned. To a statesman, this may be a guiding force in formulating a grand vision about inter-state equations depending on locations, links and flows of SLOCs that traverse a given region. Similarly, to a businessman or an economist, SLOCs may present as the most cost-effective and shortest way of transport between two given destinations.

Depending on which of these perspectives dominates a state's policies, the responses may be tilted towards either multilateral frameworks of that kind. Other initiatives are then seen as an intrusion into aspects of sovereignty and are therefore unacceptable. For instance, bigger countries in Southeast Asia - like Malaysia and Indonesia - remain opposed to India's proposals to provide protection for these regional SLOCs and the Malacca Straits in partnership with the United States. They describe this as an affront to their sovereignty while other smaller states also view such proposals with apprehension. Experience, therefore, shows how while the security of SLOCs remains premised on the success of any transnational security enforcement which requires a multilateral arrangement and cooperation, its actual working remains difficult to achieve given the acute national sensitivities, skepticism and biases, given their different priorities and historical experience.

Especially for India, given its historical experience and geographical proximity with China, and its similarly rising dependence with regard to the Indian Ocean SLOCs, China's rising focus on these SLOCs gains critical significance for India's policy-makers both as a challenge and also as a precedent to reckon with. The same, of course, remains equally true of India building a tri-Services Command in the Andamans and its initiatives to provide protection (along with the US) to these SLOCs across the Malacca Straits that so often generates unusual curiosity amongst China's power elite. The same also remains true of China's mutual perceptions with other players such as Korea and Japan, and all of this together continues to make China's maritime vision preoccupied with ensuring safety and free access to the SLOCs around the Indian Ocean rim.

References

Asia Times (Singapore), 'Kra Canal: Mega-projects anyone?', [Editorial], 28 April 2000.

Beier, Marshall J. (2005), 'Bear Facts and Dragon Boats: Rethinking the Modernization of Chinese Naval Power', *Contemporary Security Policy*, Vol. 26 (2).

Campbell, Kurt M. and Derek J. Mitchell (2001), 'Crisis in the Taiwan Strait?', *Foreign Affairs* (New York), July/August.

Chan Hok-lam (1971), 'The Chien-wen, Yung-lo, Hung-his, and Hsuan-te reigns, 1399-1435', in Frederick W. Molte and Denis Twitchett, eds., *The Cambridge History of China, Vol. 7, The Ming Dynasty, 1368-1644, Part I*, Cambridge: Cambridge University Press.

Cole, B. D. (2005), 'Waterways and Strategy: China's Priorities', The Jamestown Foundation, *China Brief*, 20 February.

Cooper, B. (2000), 'The Coming Conflict: the battle of the SLOCs', *Asia Pacific Defence Reporter*, Vol. 25 Part 8, pp. 10-12.

Dong Guozheng (2004), 'Heping jueqi nai Zhongguo de guoji zhanglue jueze' (Peaceful Rise is China's International Strategic Choice), *Jiefangjun Bao* (Liberation Army Daily), 5 April.

Fairbank, John K. (1983), 'Maritime and Continental in China's History', in John K. Fairbank, ed., *The Cambridge History of China, Vol. 12*, Republican China 1912-1949, Part I, Cambridge: Cambridge University Press.

Foot, Rosemary (1998), 'China in the ASEAN Regional Forum', *Asian Survey*, Vol. 38 (5).

Gang Deng (1999), *Maritime Sector, Institutions and Sea Power of Pre-modern China*, Westport, US: Greenwood.

Gertz, B. (2005), 'China builds up strategic sea lanes', *The Washington Times* (Washington DC), 19 January.

Hamilton, Booz Allen (2005), *Energy Futures in Asia 2004*, Office of the Net Assessment of the United States, Washington DC.

Howarth, Peter (2006), China's Rising Sea Power: *The PLA Navy's Submarine Challenge, New York*: Routledge.

Kondapalli Srikanth (2001), *China's Naval Power*, New Delhi: Knowledge World.

Kristof, Nicholas D. and Sheryl WeDunn (1994), *China Wakes: The Struggle for the Soul of Rising Power*, New York: Times Books, 1994.

Lelyveld, M. (1997), 'Greater Asian role seen in Persian Gulf, Caspian Sea', *Journal of Commerce*, 12 November, p. 20.

Radio Free Asia (2005), 'China, Philippines, Vietnam Sign Joint South China Sea Oil Search Accord', 21st March, available at:http://www.expertclick.com/ NewsReleaseWire/ default.cfm?Action=ReleasePrint nt$&ID=8473.

Military Balance (2005), *Military Balance 2004-2005*, London: International Institute of Strategic Studies.

Reynolds, Clark G. (1989), *History and the Sea: Essays on Maritime Strategies*, Columbia SC: University of South Carolina Press.

Roscini, M. (2002), 'The Navigational Rights of Nuclear Ships', *Lieden Journal* of *International Law*, pp. 255-56.

Ronan, William J. (1936), 'The Kra Canal: A Suez for Japan?', *Pacific Affairs,* Vol. 9 (3), pp. 406-412.

Ross, Robert S. (1997), 'Beijing as a Conservative Power', *Foreign Affairs*, Vol. 76 (2).

Roy, M. (2001), 'Energy and Sea Lines of Communication', *Journal of Indian Ocean Studies* (New Delhi), Vol. 9 (1).

Sakhuja, Vijay (2004), 'Chinese Navy: Strategy and Diplomacy', *Agni* (New Delhi), Vol. VII, No. IV.

Saunders, Stephen (2004), *Jane's Fighting Ships*, Surrey, UK: Jane's Publishing House.

Shambaugh, D. (1994), 'The Insecurity of Security: The PLA's Evolving Doctrine and Threat Perceptions towards 2000', *Journal of Northeast Asian Security*, Vol. 13 (1), pp. 14-15.

Sharpe, Richard, ed. (2005), *Jane's Fighting Ships 2004-2005*, Coulsdon: Jane's Information Group.

Singh, S. (2003), *China-South Asia: Issues, Equations, Policies,* New Delhi: Lancers Books.

SIPRI (2005), *SIPRI Year on Disarmament and International Security* 2005, Oxford: Oxford University Press.

Storey, Ian (2006), 'China's "Malacca Dilemma"', *China Brief,* (Jamestown Foundation), Vol. 6, Issue 8, 12 April.

Tai Ming Cheung (1990), *Growth of Chinese Naval Power,* Pacific Strategic Paper 1, Singapore: Institute of Southeast Asia Studies, pp. 38;

Taiwan Affairs Office of the State Council (2005), 'Anti-Succession Law Adopted by the NPC', Beijing, full text available at: http://www.gwytb.gov.cn:8088/detail.asp/table=headlines&title=Headlines&m_id=319.

The Age (Bangkok), 'Thai canal plan to save millions', 29 March 2005.

US Department of Defense (1995), Office of International Security Affairs, United *States Security Strategy for the East Asia-Pacific Region,* Washington DC.

Valencia, M. (1997), 'Asia, the Law of the Sea and International Relations', *International Affairs* (RUSI, London), Vol. 73 (2) pp. 263.

Wortzel, Larry M. (1994), 'China Pursues Traditional Great-Power Status', *Orbis,* Vol. 38 (2).

You Ji (1995), 'A Test Case for China's Defence and Foreign Policies', *Contemporary Southeast Asia,* Vol. 16 (4), pp. 378-380.

You Ji (2002), *The Evolution of China's Maritime Combat Doctrines and Models: 1949-2001,* Singapore: Institute of Defense and Strategies Studies.

You Ji and You Xu (1991), 'In Search of Blue Water Power: The PLA Navy's Maritime Strategy in the 1990s', *The Pacific Review,* Vol. 4 (2), pp. 137-139.

Yung, Christopher (1996), *People's War at Sea: Chinese Naval Power in the Twenty-first Century,* Alexandria, VA: Centre for Naval Analyses.

Zhang Guihong (2005), 'China's Peaceful Rise and Sino-Indian Relations', *China Report* (Delhi), Vol. 4 (2), pp. 163-164.

Chapter 5

Maritime Affairs in the Mozambique Channel

Christian Bouchard

Introduction

In the region of the South West Indian Ocean between the island of Madagascar and Mozambique on the African continent, the Mozambique Channel generally receives little attention from Indian Ocean maritime security specialists even though it represents an international waterway of some importance. This mostly results from the fact that its closure to navigation is not actually considered as a probability, nor as a very serious issue as maritime traffic could easily be maintained by bypassing Madagascar to the east. However, the Channel cannot be completely ignored, especially if we consider it to be an essential part of the Cape Route that links the Indian Ocean to the Atlantic Ocean via the south of Africa. In addition, high level of poverty on its shores, international territorial disputes, unsettled maritime boundaries, intensity of oil flow, competition for maritime resources, environmental concerns and naval operations all together create a very interesting maritime regional context that certainly deserves to be better documented.

Measuring 1,600 km in length (north-south), with a general width of 400 to 950 km (east-west), and a maximum depth of about 3,000 m, the Channel is crossed by the relatively warm Mozambique Current that flows from north to south [1] Meteorological conditions in the region are usually clement for navigation that enjoys a calm sea and good visibility. Nevertheless, from December to April, tropical storms and cyclones can temporarily complicate maritime activities into the Channel (mostly in its northern half). Located on its northern entrance, the Comoros Archipelago does not represent any difficulty for navigation as the shortest distance between the islands is 40 nautical miles (75 km) and with its depth exceeding several hundred meters.

In addition to Madagascar and Mozambique, the Union of Comoros (including the islands of Grande Comore, Anjouan and Mohéli) and France (administering the islands of Mayotte, Europa, Bassas da India, Juan de Nova and Glorioso [2] are the two other coastal states of the Channel (Figure 1 and Table 1). The land-locked States of Malawi, Swaziland, Zambia and Zimbabwe, as well as Northeast South Africa, Eastern Botswana and even southern Democratic Republic of Congo are also strongly bonded to the Channel as the Mozambican ports represent their most convenient access to and from the sea. Thus, if we consider the Mozambique Channel Maritime Region (MCMR) to be formally made up of coastal states, littoral territories and maritime spaces, the Channel hinterland extends much deeper into the southern African mainland. Since it is also a transit way for international navigation, several other states from outside the region, mostly large economic powers and Persian Gulf oil exporters, share an interest in the region, especially in terms of maritime security.

A rapid evaluation of the usual criteria used for regionalisation shows that the Mozambique Channel area clearly lacks political, cultural and economic integration, and thus does not form a real and coherent regional system (Table 2). However, the Channel itself can be considered as an original and unique maritime region. Even if it is firstly determined by the general geographical settings of the area, this maritime region is characterised by a specific hydrological feature (the Mozambique current) and is quite uniform in terms of ecological conditions. This maritime region naturally extends to the Channel shores where the coastal communities live and local maritime activities are organised, and with a more political-oriented approach, it can finally be viewed as including all of the coastal states' maritime spaces (territorial sea, contiguous zone, exclusive economic zone and continental shelf, as well as eventually archipelagic waters in the case of Comoros), with the exception of the Malagasy east coast waters. According to this definition, the Mozambique Channel Maritime Region (MCMR) covers some 1.9 million sq km of maritime spaces.

The purpose of this chapter is to discuss the general context of the Mozambique Channel Maritime Region. To do so, we will address the following issues: MCMR as part of the South West Indian Ocean geostrategic quadrant, maritime territorialisation, marine environment and resources, maritime transportation and port activities, naval presence and operations, and maritime security issues.

Figure 1
The Mozambique Channel Maritime Region

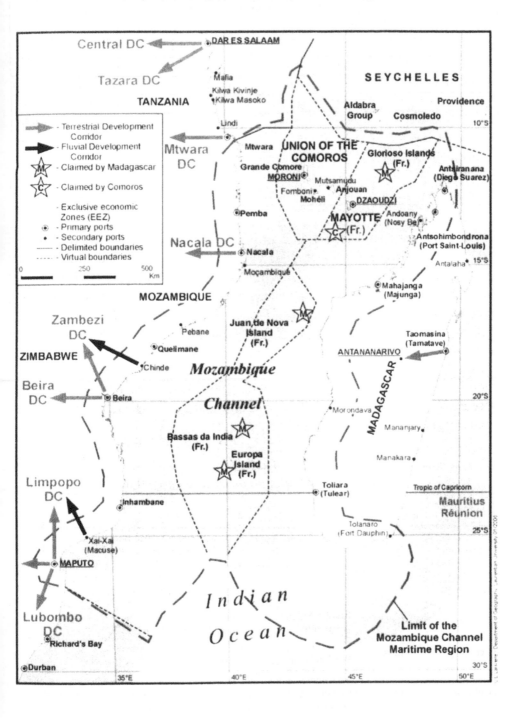

Table 1
Basic data for the coastal states and island territories of the Mozambique Channel

	Land Area (sq km)	Population mid 2005 (inhabitant)	Inhabitants per sq km	GDP[a] (million of PPP-US$)	GDP per capita (PPP-US$)	HDI[b] 2002	Claimed EEZ[c] (sq km)
Comoros	2,170	671,247	309	441	657	0.530	161,993
Madagascar	581,540	18,040,341	31	14,560	807	0.469	1,079,612
Mayotte (France)	374	193,633	518	467	2412	[0.600]	50,000
Mozambique	784,090	19,406,703	25	23,380	1205	0.354	493,672
Scattered Is. (France)[d]	39.6	0	-----	-----	-----	-----	360,400
- Bassasda India	0.2	0	-----	-----	-----	-----	123,700
- Europa Island	28	0	-----	-----	-----	-----	127,300
- Gloriosos Is.	7	0	-----	-----	-----	-----	48,350
- Juan da Nova	4.4	0	-----	-----	-----	-----	61,050

(a) Gross domestic product in million of purchasing power parity US$ (PPP-US$): 2002 for Comoros, 2003 for Mayotte, 2004 for Madagascar and Mozambique; Gross domestic product per capita in purchasing power parity US$ (PPP-US$). (b) Human Development Index (estimation for Mayotte). (c) Claimed exclusive economic zone (estimations). (d) The French Sparse Islands of the Indian Ocean also include Tromelin Island (east of Madagascar). There is no local permanent population on these islands.

Main sources: CIA, The World Factbook 2005 (land area, population, GDP); UNDP, Human Development Report 2004 (HDI); WRI, EarthTrends: The Environmental Information Portal (claimed EEZ).

Table 2
Mozambique Channel as a regional system

Dimension	General appreciation	Comments
Geography	High	Waters between Madagascar and southern African mainland Waters and littoral areas of the coastal states and island territories Coastal state and island territories maritime spaces (TS, CZ, EEZ, CS)
Hydrology	High	Mozambique current
Ecology	High	Eastern African Marine Ecoregion (EAME) Western Indian Ocean Marine Ecoregion (WIOME)
Economy	Low	No regional system between coastal state economies Weak relation between coastal communities of the two sides
Cultural	Low	No common culture and identity: Swahili in Northern Mozambique and Comoros Archipelago, several other ethnic groups in Moz. and Mad. Distinct colonial heritage from French and Portuguese administrations
Geopolitics	Low	No specific regional association between coastal states Not seen as a unique geopolitical/geostrategic area from outside

MCMR as part of the SWIO geostrategic quadrant

In terms of geopolitics, the Mozambique Channel area does not form a specific region on its own. It is better understood as part of the Greater Southern Africa Region, which roughly covers the Southern African Development Community (SADC) member countries and the other South West Indian Ocean Islands [3]. In a more maritime-oriented context, the Channel is usually considered as part of the Western Indian Ocean or the South West Indian Ocean. Our analysis of the Indian Ocean geostrategic theatre shows that this Ocean can be divided into four specific quadrants converging together at Diego Garcia from where the Americans can project enormous power and military support in any direction (Figure 2). This division seems very logical and meaningful as the North West Quadrant is linked to the Greater Middle East Region, the North East Quadrant is linked to the Bay of Bengal/Malacca Strait Region, the South East Quadrant is linked to the Australia-Indonesia Area, and the South West Quadrant is linked to the Greater Southern Africa Region.

The South West Quadrant's main features are the maritime traffic related to the Cape Route, the regional preeminence of the French Navy and some good fishing grounds. The Mozambique Channel is only one of several Indian Ocean routes linked to the Cape Route that circles around the southern tip of Africa and connects with the Atlantic Ocean. Recognised as one of the four 'doors' of the Indian Ocean, the Cape Route is generally considered of less importance than those of the Malacca/Indonesian Straits and of the Suez Canal/Red Sea, but of greater significance than the Australian Straits (Torres and Bass). If the maritime traffic in the Southwest Indian Ocean door is effectively less significant than it is in both the Northeast and Northwest doors, its strategic weight is enhanced by the fact that the channel represents the back-up solution to reach Europe and the Eastern coasts of America from the Indian Ocean in case of a closure of the Suez Canal. As this situation recently happened twice in the near past (1956-57 and 1967-75), and as the Middle East is still very volatile, another closure cannot be totally disregarded.

Figure 2
The Four Indian Ocean Geostrategic Quadrants

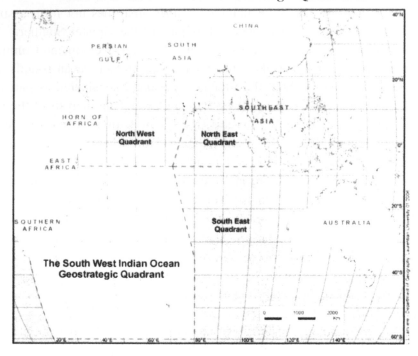

Figure 3
The Southwest Indian Ocean Geostrategic Quadrant

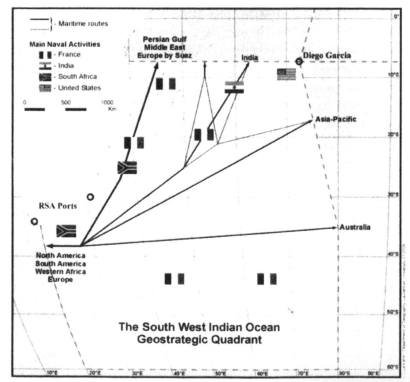

In terms of naval presence, the South West Indian Ocean Quadrant is mostly French-dominated since no other large power has been continuously operating a wide range of operations in this area. In addition to the French, the South African Navy is patrolling in the Mozambique Channel and in the southern waters, while India has lately showed great interest in the region and has begun to operate on a permanent basis in the Mauritian exclusive economic zone. In addition, the Indian Navy was in charge of the maritime security of the African Summit held in Mozambique in 2003 and held exercises with the South African and French navies. If the Americans are mostly relying on the French to monitor and control this maritime area, they have been developing relations with regional governments and have directed some of their attention to the area in the context of the global war on terrorism. US Navy vessels, including combat groups and Military Sealift Command ships, are frequently sailing through the region on their way to or from the North West Quadrant or Diego Garcia.

Regarding marine resources exploitation, fishing is especially good in Mozambique and Madagascar coastal waters, on the Seychelles Plateau and in the cold waters surrounding the southern islands (French three districts of the Territory of the French Southern and Antarctic Lands: Crozet, Kerguelen, Saint-Paul and Amsterdam; South African Prince Edward Islands). As the regional states do not possess a fishing fleet necessary to exploit the living resources of their exclusive economic zone, foreign fishing vessels are numerous and operate in the area. These ships mostly come from the European Union, Japan, South Korea, Taiwan, and even South America (Chile, Argentina and Uruguay).

Maritime Territorialisation in the MCMR

Maritime territorialisation can be defined as the process by which coastal states' sovereignty, exclusive rights and specific jurisdictions are spatially implemented at sea (the creation of maritime zones, limits and boundaries). The international legal basis for this process is the United Nations Convention on the Law of the Sea (UNCLOS) that was adopted in 1982 and entered into force in 1994. In regard to territorialisation, the Convention defines the general principles for setting national baselines and delimiting the different maritime zones where specific legal regimes apply. In this context, each coastal state can, where the general geographical configuration make it possible, extend its sovereignty to a zone defined as the territorial sea (up to 12 nautical miles from the baselines), implement some specific jurisdictions into a zone defined as the continuous zone (from the territorial sea limit and up to 24 nautical miles from the baselines), and exercise some exclusive rights on marine resources in zones defined as the exclusive economic zone (from the territorial sea limit and up to 200 nautical miles from the baselines) and the continental shelf (up to the external limit of the continental shelf or to 200 nautical miles from the baselines where this limit is closer to 200 miles). Even if UNCLOS provides the basic principles, the process of maritime territorialisation is largely a state affair as each coastal state has to set its own baselines, create the necessary legislation to implement the different maritime regimes and negotiate maritime boundaries with its neighbours.

Table 3
Claims to Maritime Jurisdiction in the Mozambique Channel
(in nautical miles)

	UNCLOS ratification	Territorial Sea	Contiguous Zone	Exclusive Economic Zone	Continental Shelf	Other claim
)s	1994	12	---	200	N/A	Archipelagic status [a]
Mayotte	1996	12	24	200	200/Expl. [c]	---
- Scattered Is. [b]	1996	12	24	200	200/Expl. [c]	---
ıscar	2001	12	24	200	200/Del. [d]	---
bique	1997	12	24	200	CM/200 [e]	---

(a) But no archipelagic baseline has been provided yet. (b) The French Indian Ocean Scattered Islands include Bassas da India, Europa, Gloriosos Is., Juan de Nova in the Mozambique Channel as well as Tromelin (east of Madagascar). (c) 200 nautical miles or depth of exploitability. (d) 200 nautical miles or delimitation agreements or 100 nautical miles from isobath 2,500 m. (e) Outer edge of the continental margin, or to 200 nautical miles where the outer edge does not extend up to that distance.

Main source: DOALOS (2005), *Table of claims to maritime jurisdiction.*

Table 4
Territorial Disputes in the Mozambique Channel Maritime Region

	Administrated by	Status	Claimed by
Mayotte	France	Departmental collectivity	Comoros
Europa	France	Dependancy	Madagascar
Bassa da India	France	Dependancy	Madagascar
Juan da Nova	France	Dependancy	Madagascar
Gloriosos Is.	France	Dependancy	Madagascar [a]

(a) In 1980, Comoros declared that it will officially claim the Glorioso Is. when it will have regained Mayotte.

In the Mozambique Channel, the process of maritime territorialisation is far from complete. Even if all of the coastal states have established some maritime regulations and zones (Table 3), only two peripheral maritime boundaries have been officially set up - those between Mozambique and Tanzania (agreement concluded in 1988) and between France for Glorioso Is. and Seychelles (agreement concluded in 2001). This is mainly due to the fact that the sovereignty on all of the five French islands in the Channel remains disputed (Table 4). As the process of maritime territorialisation is land-based, sovereignty issues usually have to be resolved before maritime boundaries can be agreed upon [4]. Regarding both Mayotte and the four Scattered Islands, respectively claimed by Comoros and Madagascar, the latest developments show that France has definitively made the choice of maintaining its sovereignty over them.

In the case of the Scattered Is., France has accepted in December 1999 the principle of common management of the disputed territories with its counterparts in the Indian Ocean Commission (IOC), but there has been no official development since. Some say that there is nothing to manage because the Scattered Is. have been designated as a natural reserve since 1975, but it is the exclusive economic zones and their marine resources that are at stake, much more than the small islands themselves. In the Mozambique Channel Maritime Region, France claims some 360,000 sq km of EEZ around the four concerned Scattered Is. (Table 5). In this area, common management of marine resources exploitation could eventually be agreed upon between France and Madagascar, even if the sovereignty issue remains unresolved.

Regarding Mayotte, the situation has evolved substantially since the beginning of the 1990s. France has slowly but finally accepted to fully commit itself to the socioeconomic development of the island. As a result, the island obtained a new political status in 2000 that deepens its relation with France and puts it on the way to complete national integration if desired by the local inhabitants (possibility of eventually gaining the status of French Overseas Department and thus of integrating into the European Union territory). Facing this new reality, the Comoros claim over Mayotte appears to be futile even if its arguments on the fundamentals of the dispute are not necessarily illegitimate. The problem here is that the Moharais themselves are strongly opposed to return to Comoros sovereignty and claim their rights to self-determination (a principle that France has fully recognised). In this context, Comoros could probably attempt to claim compensation form France for the loss of Mayotte as part of an eventual settlement of the dispute. This dispute has finally a small effect on the maritime territorialisation of the Mozambique Channel, but remains a very important issue in regard to regional cooperation among the South West Indian Ocean Islands.

Actually, France is a member of the Indian Ocean Commission (IOC) but it can only represent Réunion Island, as the other island states (Madagascar, Mauritius and Seychelles) support the Comoros position regarding its claim over Mayotte. Thus, Mayotte is politically quite isolated both from it closest neighbour (Comoros) and from the regional island community (IOC).

Table 5
Estimated Maritime Spaces of the MCMR Coastal States and Island Territories

	Land area (sq km)	Length of coastline (km)	Territorial sea (sq km)	Exclusive Econ. Zone (sq km)	Continental Shelf [a] (sq km)	EEZ to land ratio	EEZ to coastline ratio
Comoros	2,170	469	12,684	161,993	1,416	75	345
France - Mayotte	374	185.2	N/A	50,000	N/A	134	270
- Bassa da India	0.2	35.2	N/A	123,700	N/A	618,500	3,514
France - Europa	28	22.2	N/A	127,300	N/A	4,546	5,734
France – Gloriosos Is.	5.0	35.2	N/A	48,350	N/A	9,670	1,374
France – Juan da Nova	4.4	24.1	N/A	61,050	N/A	13,875	2,533
Madagascar [b]	581,540	9,935	124,938	1,079,672	96,653	1.9	109
Mozambique	784,090	6,942	70,894	493,672	73,300	0.6	71

(a) Up to 200 meters depth. (b) About 60% of Madagascar coastline is lying on the Mozambique Channel. N/A: not available.

Main sources: WRI, EarthTrends (for Comoros, Madagascar and Mozambique); CIA, *The World Factbook 2005* (for French islands land area and length of coastline); France, Ministère de l'Outre-Mer (for French islands claimed EEZ).

Another issue regarding maritime territorialisation is that of the Comoros claim for archipelagic status. In accordance with UNCLOS principles [5], Comoros should now determine its archipelagic baselines from which its territorial sea and its other maritime zones can be drawn. Inside these baselines, the legal regime will be that of the archipelagic waters on which Comoros will be able to exercise its sovereignty. Finally, as all of the Mozambique Channel waters fall into the 200 nautical mile limit of the coastal states, international maritime boundaries will have to be negotiated to fix all of the exclusive economic zone and continental shelf outer limits.

Marine Environment and Resources

The Mozambique Channel is essentially a warm tropical body of water characterised by the Mozambique current that runs from north to south, originating north of the Comoros and Madagascar from the South Equatorial current (that crosses the Indian Ocean from east to west) and meeting southward with the Agulhas current that swings around from Southern Madagascar and runs south along the Kwa-Zulu-Natal coast of South Africa. According to the WWF's Ecoregion Programme (Global 2000), the Channel waters relate to two sister Ecoregions which are the Eastern African Marine Ecoregion, stretching along the African coast from Southern Somalia to the Natal shores of South Africa, and the Western Indian Ocean Ecoregion, covering the waters surrounding the South West Indian Ocean islands. In the Channel, there is no reason to make any difference between both Ecoregions as the environmental conditions and marine species are quite constant over the whole area.

The main marine habitats are mangrove forests, sea grass beds, coral reefs and open water of generally great depth as the continental shelf is usually narrow on both Mozambican and Malagasy sides of the Channel. In terms of particular marine animals, these waters are renowned for the presence of the coelacanth, dugong, and several species of dolphins, sea turtles, sharks and whales. Several sites of global and regional importance have been identified and are considered as priorities for conservation. In addition to the Mozambican sites (see WWF, 2005), the list includes the four French Scattered Islands that are already designated as natural reserves, the lagoon of Mayotte (one of the world's largest with an area of about 1,000 sq km), Aldabra atoll (on the UNESCO World Heritage list), and by several Madagascar sites. In this latter case, the state is now expanding its conservation effort to include the marine environment after several decades of prioritising forest and land habitats.

Regarding fisheries, the catches are not very important in terms of global quantities even if many coastal communities are relying solely or mainly on fishing for both their food supplies and their economy. Local artisan fisheries are generally limited to coastal waters where they are often competing with larger commercial boats for shrimps, which represent by far the most important commercial resources of these waters. Shrimp and prawn culture are also developed on the Mozambican littoral. Distant waters are mainly fished by foreign ships from Europe (France and Spain) and Asia (South Korea, Taiwan, Japan). The main catches in these waters are tuna species. The last FAO report on the state of the world fisheries and aquaculture (2004) concludes for the Western Indian Ocean that some 75% of the fishing resources are fully developed while 25% are actually overexploited. The particular situation in the Mozambique Channel is not much better while some coastal fisheries have already declined and distant fisheries have remained quite stable over the last few years. The development of aquaculture (mostly shrimps) is seen as a real opportunity to increase production and export revenue in both Mozambique and Madagascar.

Regarding minerals resources, oil and gas are certainly the main issues in the Channel. Hopes for a serious oil discovery lasted for several decades in this area, but were tempered by the cost of exploration and exploitation techniques in an area of relatively deep water (depth averaging around 2,000 m). It is in this context that an extensive geologic and geophysical study of the Mozambique Channel has recently been completed by Rusk, Bertagne and Associates in collaboration with TGS-NOPEC (Rusk, Bertagne and Associates, 2003). The study, which has been conducted with the objective of establishing the real potential for oil in the area, indicates the presence of "multi-petroleum systems combined with mega-anticlinal structures" (Shirley, 2003). "Although major gas reserves (3.5 TCF) of the Pande-Temane complex in Mozambique, and the Bemolanga tar sands (20 billion barrels) and Tsimiroro oilfield (8 billion barrels) in Madagascar have been known for years, a total of only 23 exploration wells have been drilled offshore" (Rusk, Bertagne and Associates, 2003). This evidence for a major oil discovery has resulted in serious interest in the region and several new exploration campaigns are now underway. For the authors of the study, the regional oil potential should be at least comparable to that of the North Sea and maybe more as the basic structures exhibit Middle East reserves potential.

Thus, in the next few decades, major oil developments could modify quite seriously the whole maritime context of the Mozambique Channel. If the environmental and economic impacts are regionally important, the global geostrategic importance of the Channel will also significantly increase. Other marine resources such as salt or marine energy (from waves, tides or thermal gradient) are neither viewed today, in the region, as important (salt) or very promising resources (marine energy).

Finally, regarding the environment and living resources, the problem of coherent and collective management remains even if several regional inter-state institutions are now operating in this field. If there is no specific Mozambique Channel international body, Channel issues are worked out in a wider geographic context by several institutions of which the main are: the Eastern African Region - UNEP Regional Seas (EAF-UNEP-RSC), the Indian Ocean Tuna Commission (IOTC), the South West Indian Ocean Fisheries Commission (SWIOFC), the South West Indian Ocean Fisheries Project (SWIOFP) and the Indian Ocean Commission (IOC).

Maritime Transportation and Port Activities

In regard to maritime transportation, the Mozambique Channel is essentially viewed as an international transit area as the regional activities and interactions are quite weak. In terms of legal regime, ships crossing the Channel enjoy freedom of navigation in all the waters that extend outside of the coastal states territorial seas. There is only between Grande Comore and Anjouan in the Comoros Archipelago that the whole passage is submitted to the territorial sea regime of navigation that is of innocent passage. This could eventually evolve if Comoros officially defines its archipelagic waters.

In terms of intensity of transit, the number of ships remains low compared to the busiest world choke points. Nevertheless, according to the International Tanker Owners Pollution Federation Limited (ITOPF, 2003), "a major tanker route stretches along the East African coast, from the Middle East to Europe or United States. Around 5,000 tanker voyages per year are reported, carrying 30% of the world's crude oil production. Of these, 1200 voyages per year are by large tanker (>250,000 tonnes) and 4,000 are by middle-sized tanker. Of the 700 million tones per year of crude oil sailing in the island waters of the Indian Ocean, 350 million tones are transported through the Mozambique Channel". Besides oil tankers, the Channel is of lesser importance as most of the container and general cargo ships traveling into the Indian Ocean by the Cape Route do not follow the African coast but instead navigate straight to Australia and Southeast Asia, with some stopping either in Mauritius (Port Louis) or Réunion (Le Port). Even the MSC Indian Ocean Service that links Durban, Singapore and Fremantle with secondary connections does not pass through the Channel.

The maritime regional network is largely organised around the port of Durban that serves as the main, not to say the unique, transshipment platform for the whole Mozambique Channel. From there, cargoes are dispatched or collected by regional services that go as far as Tanzania, Kenya and Seychelles to the North and to Taomasina (Tamatave), Réunion and Mauritius to the East (Figures 1 and 3). In this regional framework, Port Louis and Port Réunion (Le Port) are also becoming regional hubs for cargo transiting to and from Madagascar, Comoros, Mayotte and Seychelles. In the Channel itself, the Malagasy ports remain of minor importance as both the local cargoes and their hinterlands are small. The situation is very different on the Mozambican coast, where, if the local cargo remains small (except for Maputo which exports several bulk products), the port's hinterlands are very wide. Maputo, Beira and Nacala are effectively playing an important transit role for the land-locked regions of southeastern Africa. The rail, road and river systems that link these ports to the interior of the continent are the backbones of integrated development initiatives known as development corridors (Figure 1). For now, both the quantity of cargo in transit and the industrial development have been very moderate in these corridors, but infrastructure modernization, new and improved management and regional coordination should eventually contribute to a rapid growth of the activities in the Mozambican regional ports. Considering the congestion of the ports of Durban and Port Elizabeth, the port of Maputo can certainly become a port of primary importance for South Africa's economic heart (the Johannesburg-Pretoria area) to which it is already quite well linked by terrestrial transportation.

Naval Presence and Operations

The Mozambique Channel was one of the hot spots of Indian Ocean naval competition that occurred in the Cold War. At that time, the Soviet Union maintained submarines and some surface vessels in the area with a view to disrupting the oil flow from the Middle East to the United States, if necessary. At the same time, France was confirming its willingness to play a key military role in this area of the world and was maintaining a strong western Indian Ocean naval presence that relied on its naval facilities in Djibouti and Réunion. Sharing information with the Americans, France was granted the mission of surveying the South West Indian Ocean waters for the Western Free World. If the Soviets are gone, the French Navy is still there and has carried on with its permanent presence in the area. It can certainly be said that the French Navy was almost the only important player in the Channel for all of the 1980s and 1990s. However, the situation is now evolving rapidly as the South African Navy is modernised and intends to expand its operations at the southern Africa scale, the Indian Navy is showing interest and the flag in the Channel as well as the US Navy combat and Maritime Sealift Command ships which are cruising more and more in the region on their way to or from the Middle East and Diego Garcia (Table 6).

For now, the whole South West Indian Ocean Geostrategic Quadrant remains certainly a French-dominated area, but the French now have to share these waters with newcomers. Regarding the Mozambique Channel, the important relation that the French and South African Navies have developed is certainly very positive for the region. This means that the increasing regional role for the South African Navy is happening in a context of collaboration and not of confrontation; collaboration that extends to Mozambique and to Madagascar.

Table 6
Major Navies Operating in the Mozambique Channel

States	Intensity of the presence	Main military operations	Base or facility
France	Permanent	Sovereignty, EEZ patrol, Traffic surveillance, Intelligence, Exercise, Port visit, Humanitarian Relief, Special OP Military operations in Comoros: 1989, 1992, 1995 Humanitarian relief in Madagascar (Cyclone) Frequent joint exercises with the SA Navy and Indian Navy	Several [a]
South Africa	Permanent (?)	Traffic surveillance, Intelligence, Exercise, Port visit New patrol corvettes/frigates (4)and submarines (3) Frequent joint exercises and operations with the France Provided (with France) armed patrol boats to Moz. (2004)	Very close
United States	Frequent	Transit for battle groups, pre-positioning and cargo ships, Port visit, Intelligence (?) At the edge of the PACOM, EUCOM, CENTCOM, and CJTF-HOA areas of responsibility Provided 6+1 motor lifeboats to Madagascar (2003)	Distant
India	Sporadic	Special OP, Exercise, Port visit Security OP: African Union Summit in Mozambique (2003) Naval visit to Cape Town and Durban (4 ships, June 2005)	Distant

(a) Including port facilities and aerial support in Mayotte, as well as small airfields on the Scattered Islands.

Maritime Security Issues

In regard to maritime security issues, the Mozambique Channel Maritime Region is not internationally recognised as a very sensitive area. Nevertheless, our own preliminary and very basic evaluation of the risks reveal that this particular maritime area experiences a wide range of threats, several of which are considered to be very serious (Table 7). In general, risks related to the international maritime transportation sector are relatively low while those related to local illegal activities are much more important. The former case is related to the easy navigational environment in which large commercial ships operate and to the relatively low traffic density in the area. The latter case is related to the fact that the Channel coastal and distant waters remain quite poorly surveyed as coastal states only possess very limited maritime patrol capabilities. To help the coastal states address this problem, both Mozambique and Madagascar have received new coastal patrol boats donated by South Africa and France to Mozambique in 2004 and by the United States to Madagascar in 2003. France is also providing from time to time some direct assistance to Comoros and Madagascar in patrolling in their exclusive economic zones.

Table 7

Main maritime security issues in the Mozambique Channel Area

	General level of risk	Comments
Ship collision	Low	Wide space for navigation, low intensity of the regional traffic
Oil spell	Medium	High number of oil tankers, several VLCC
Port security	Medium to High	Concerns about cargo handling and security Cargo in transit to or from the land-locked states
Terrorism	Medium to High	From Tanzania or Comoros
Piracy	Low	No organized group operating in the area (for now)
Robbery	Medium	Some concerns close to shore, especially for small private boats
Illegal trafficking	High	Especially for drugs (Mozambique) and general goods (Comoros)
Illegal migration	Medium	Only high from Comoros and Madagascar to Mayotte
Sensitive vessels	High	Oil tankers US Navy Ships and armament/ammunition cargoes
Illegal fishing	High	From foreign ships in poorly surveyed local EEZ
Foreign navies	High	Especially for intelligence gathering and eventually for sea-to-land intervention (ex.: Comoros in 1995)

If the global situation remains quite acceptable for now, three factors can contribute to a degradation of the overall security level in which maritime activities take place actually in the Mozambique Channel. The first one is the low human development and high level of poverty in the coastal communities. This constitutes a favourable environment for increasing criminal activities not only on land but also at sea. If the large vessels cruising far offshore are relatively protected from the emerging low-level criminality, the smaller boats navigating close to the shores and visiting the local ports are particularly vulnerable. The second emerging factor of insecurity is the attested presence in Kenya, Tanzania and even Comoros of jihadism sympathisers and Al-Qeada-related groups. With the important number of large tankers operating in the area, the northern entrance of the Channel could eventually be targeted for a maritime terrorism operation. Local ports certainly also represent vulnerable infrastructures that need to be securitised. The third factor that should be considered as a challenge to the global maritime security of the region is the future oil activities. The increasing number of tankers cruising in the area and the extraction operations could certainly have negative environmental impacts in the Channel. Major oil activities can also lead to some local instabilities (like for example in the Niger Delta), fuelled either by local discontentment in poor coastal communities or by outside actors (environmentalists/oil industry opponents, opportunist political groups, terrorists, and so on).

Conclusion

Far from the Persian Gulf and the Straits of Malacca, the Mozambique Channel is often forgotten by geopoliticians and geostrategisits working on Indian Ocean issues. Nevertheless, our introductory work shows that its maritime context is very rich in terms of local, regional and international issues. This context is especially characterised by unsettled territorial disputes and a maritime territorialisation process that is far from complete, coastal environmental concerns and great pressure on all the living resources of the sea, a serious potential for large offshore oil discoveries, transit of an important number of oil tankers (VLCC and medium-sized tankers), navy newcomers in this traditionally French-controlled area, as well as by a wide range of maritime security issues.

If additional research work needs to be done to give a more comprehensive and specific evaluation of the maritime security issues in the Mozambique Channel Maritime Region, there is certainly no doubt that it is emerging as an area of more intense competition for maritime resources (including living resources as well as offshore oil and gas) and of raising security concerns for maritime transportation (including ship transit and port activities). In terms of security at sea, the context is evolving quite rapidly with the development of both local capacities (navy and coast guard) and foreign naval presence (US and India joining France and South Africa). As for outside players getting seriously involved in the region, China is certainly also a consideration as it has clearly shown interest in deepening relations with Madagascar, Mozambique and South Africa (with whom it met in November 2005 to discuss strengthening military cooperation). This renewed attention of the great foreign powers for the area clearly shows that the Mozambique Channel Maritime Region and, more generally, the South West Indian Ocean are definitively becoming more closely linked to the world economy and to the scene of world affairs.

Endnotes

(1) "The southeast trade winds move the Indian South Equatorial Current toward the east coast of Africa, off which, because of the Earth's rotation, it is directed south to follow the outline of the mainland and its continental shelf. While some of this flow passes east of the island of Madagascar, the rest funnels to the west through the Mozambique Channel, bringing with it strong influences on the climate of the island and mainland. South of Madagascar both streams feed into the Agulhas Current." (Encyclopedia Britannica, Internet)

(2) Europa, Bassas da India, Juan de Nova and Glorioso Is., along with Tromelin that lies east of Madagascar, are usually referred to as the Scattered Islands (oles eparses), or the French Indian Ocean Scattered Islands. These islands have no permanent population and are administered by the Superior administrator of the French Southern and Antarctic Lands (located in Reunion Island).

(3) SADC members are Angola, Botswana, Democratic Republic of Congo, Lesotho, Malawi, Mauritius, Mozambique Namibia, South Africa, Swaziland, United Republic of Tanzania, Zambia and Zimbabwe. The other South West Indian Ocean islands are Comoros, Madagascar, Mayotte (France), Reunion (France) and Seychelles.

(4) Two States can also decide to freeze the sovereignty issue and agree on a common management of the disputed area.

(5) "An archipelagic State may draw straight archipelagic baselines joining the outermost points of the outermost islands and drying reefs of the archipelago provided that within such baselines are included the main islands and an area in which the ratio of the area of the water to the area of the land, including atolls, is between 1 to 1 and 9 to 1" (UNCLOS, Point 1, Article 47).

Acknowledgement

Special thanks to M. Léo L. Larivière, Technologist for the Department of Geography at Laurentian University (Sudbury, Canada), for the realisation of the maps.

References

CIA (2005), *The World Factbook 2005*. Central Information Agency, Washington. On the Web at: http://www.cia.gov/cia/publications/factbook/index.html (last visited on June 19th, 2005).

DOALOS (2005), *Table of claims to maritime jurisdiction*. New York: United Nations, Division for Ocean Affairs and the Law of the Sea (DOALOS). On the Web at: http://www.un.org/Depts/los/LEGISLATIONANDTREATIES/PDFFILES/claims_2005.pdf (last visited on May 18th, 2006).

EAF / UNEP-RSC, Internet: The Eastern African Region - UNEP Regional Seas Webpage. On the Web at: http://hq.unep.org/easternafrica/ (last visited on June 20th, 2005).

FAO (1997), *Review of the state of world fishery resources: Marine fisheries*. Rome: Food and Agriculture Organization of the United Nations, FAO Fisheries Circulars - C920. On the Web at: http://www.fao.org/documents/ show_cdr.asp?url_file=/DOCREP/003/W4248E/W4248E00.HTM   (last visited on June 27th, 2005).

FAO (2004), *The State of World Fisheries and Aquaculture 2004*. Rome: Food and Agriculture Organization of the United Nations, 153 pages. On the Web at: http://www.fao.org/documents/show_cdr.asp?url_file=/DOCREP/007/y5600e/y5600e00.htm (last visited on June 27th, 2005).

France, Internet: " Presentation des Oles eparses ", *Ministere de l'Outre-Mer Webpage*. On the Web at: http://www.outre-mer.gouv.fr/outremer/front (last visited on June 19th, 2005).

IOC, Internet: The Indian Ocean Commission Webpage. On the Web at: http://www.coi-info.org/

IOTC, Internet: The Indian Ocean Tuna Commission Webpage. On the Web at: http://www.iotc.org/

Rusk, Bertagne and Associates (2003), Petroleum geology and geophysics of the Mozambique Channel. Executive summary available on Internet (last visited on June 27th, 2005).

Shirley, Kathy (2003), 'Cold war maps prove boon for explorers: big East Africa Structures enticing', AAPG: Explorer, July 2003. On the Web at: http://www.aapg.org/explorer/2003/07jul/eastafrica.cfm (last visited on June 20th, 2005).

SWIOFC, Internet: The South West Indian Ocean Fisheries Commission Webpage. On the Web at: http://www.fao.org/fi/body/rfb/SWIOFC/swiofc_home.htm (last visited on June 20th, 2005).

SWIOFP, Internet: The South West Indian Ocean Fisheries Project Webpage. On the Web at: http://www.swiofp.org/ (last visited on June 20th, 2005).

WRI, Internet: EarthTrends: *The Environmental Information Portal. Washington* DC: World Resources Institute. On the Web at: http://earthtrends.wri.org (last visited on June 19th, 2005).

WWF (2005), *East Africa Marine Ecoregion Brochure.* Dar es Salaam: World Wildlife Fund Tanzania Programme Office. On the Web at: http://www.panda.org/downloads/eamebrochureelectronic.pdf (last visited on May 18th, 2006).

THE INDIAN OCEAN AS A GLOBAL TRADING AND ENERGY LIFELINE

Chapter 6

The Role of the Indian Ocean in Facilitating Global Maritime Trade

Nazery Khalid

This chapter explores the role of the Indian Ocean in promoting global maritime trade by providing an assessment of the importance of this waterway to world commerce. The chapter examines several factors contributing to the ocean's prominence as a crucial trade waterway and discusses its importance as a facilitator of international commerce from the context of the development of ports and the shipping sector in the region. It also analyses the prospects for the further growth in maritime trade in the Indian Ocean and its region.

Indian Ocean: Vital Waters, Critical Area

The Indian Ocean (IO) is the world's third largest ocean, covering 20% of the earth's water surface (Wikipedia web site, 2005). It is a major sea lane connecting the Middle East, East Asia and Africa with Europe and the Americas. Boasting rich living and non-living resources, from marine life to oil and natural gas, the IO is economically crucial to Africa, Asia and Australasia, the three continents bordering it, and the world at large.

The IO is a critical waterway for global trade and commerce. This strategic expanse hosts heavy international maritime traffic that includes half of the world's containerised cargo, one third of its bulk cargo and two thirds of its oil shipments (Sakhuja, 2003). Its waters carry a heavy traffic of petroleum and petroleum products from the oilfields of the Persian Gulf and Indonesia, and contain an estimated 40% of the world's offshore oil production (CIA web site, 2005). In addition to providing precious minerals and energy resources, the ocean's fish are of great importance to the littoral states for domestic consumption and export. The IO rim defines a distinctive area of much diversity in climate, culture, race, religion, language, political orientation, economic development, and strategic interest. There are approximately 40 countries either littoral to or are island states in the Ocean, including those with coastlines on bodies of water such as the Red Sea, the Persian Gulf and the Straits of Malacca. Additionally, there are about a dozen land-locked countries which have their main ocean access through the IO (Curtin University IOC web site, 2005). A number of sub-regions are evident in the IO - for example, Southern and Eastern Africa, the Horn of Africa and the Red Sea, South Asia, Southeast Asia, and Australasia. It also includes a number of regional organizations such as ASEAN, GCC, SAARC and SADEC.

Recent developments in the IO region - encompassing economy, politics and policies - point towards greater integration in the area. In an increasingly globalised world, the IO region is likely to benefit from the free flow of resources, goods and investments that can boost trade in the vicinity. The World Bank estimated that global maritime trade was expected to increase from 21,480 billion ton-miles in 1999 to 35,000 billion ton-miles in 2010, a significant percentage of which would pass through the IO (Sakhuja, 2003). Such a projection points towards growth in consumer-based industries worldwide, which, in turn, would generate greater demand for energy resources from the IO region. This is poised to enhance the Ocean's importance as a facilitator of global maritime trade.

An Overview of the IO Regional Economies

The IO region covers a large swathe of territory extending across three continents, and featuring a wide array of countries and economies. The Ocean's territory contains some of the world's most impoverished states as well as some of its richest. As such, labeling all the states in the area as mere "IO states" is only practical in a general discourse, but would not do any justice to the region's stunning diversity and disparity and to the complexity of the issues therein. Furthermore, it emphasises, once again, that regionalism is a construction. Nonetheless, for the purposes of providing an economic overview, this section will broadly highlight the extensive economic diversity of the IO region.

In a nutshell, the IO region is made of countries ranging from underdeveloped to developed. Underlining the vastness of the area and the economic contrasts among its states, there are tiny island states in the western IO possessing a weak economic base (UNEP, 1999) and there are some of the world's most prosperous countries across the region's expanse. Australia and India alone accounted for nearly 40% of total GDP in the IO region, while these two countries along with Indonesia, Iran, Malaysia, Pakistan, Saudi Arabia, Singapore, South Africa and Thailand made up around 90% of GDP in the entire region (IFIOR, 1995). Nevertheless, while several countries in the IO rim remain economically impoverished, many are becoming globally competitive and are developing new capacities, some jointly harnessed through regional cooperation efforts.

The Indian Ocean Rim Association for Regional Cooperation (IOR-ARC) was formed in 1997 with the objective of promoting economic cooperation among countries of the Indian Ocean Rim. It is the only organization of countries along the IO rim that promotes regional economic cooperation. The grouping advocates principles of open regionalism and inclusivity of membership, with the objective of encouraging trade liberalisation and cooperation. It also promotes activities focusing on trade facilitation, promotion and liberalization, investment and economic cooperation in the IO region. It is indeed a matter of debate as to the success of this regionalism initiative and it is probably fair to conclude that the relative weakness of the 18-state grouping is, in part, a reflection of the relatively weak economic ties among member states. As has been pointed out, regionalism tends to follow economic linkages rather than create them (Rumley and Chaturvedi, 2004, pp. 26).

The Indian Ocean as a Facilitator of Global Maritime Trade

The IO region has entered into a new millennium full of inspiration, promises and opportunities, along with challenges, risks and uncertainties. With these looming on the horizon, in a world increasingly dependent on foreign trade and the maritime sector to facilitate global economic growth, the IO as one of the world's most important waterways will continue to be thrust into the spotlight. The waters of this expansive Ocean stand to be intensely used in maritime transport and to facilitate international commerce.

Over the decades, the IO has become a chessboard of strategic rivalry and maneuverings of great powers, but has also played a central role in facilitating world trade and economic growth. What are the major factors that make the IO such a decisive facilitator of global maritime trade? Some essential elements identified and described in the literature on this great body of water include historical influences, population, location, trade and economic patterns, and political development, among others. To critically evaluate the importance of the IO's role as a catalyst in global maritime trade, an analysis of the major factors, and the trade and economic links between the IO rim and other regions, is warranted. The IO region is linked together by trade routes harking back from time immemorial and provides crucial access to the major sea lanes. The area is the home of the world's first urban civilisation and the centre of the first sophisticated commercial and maritime activities (Kamat, 1998). The routes through Aden and Hormuz have been used since antiquity for the purposes of trade and communication.

Today, the IO region is one of the busiest waterways in the world for the commercial exchange of commodities, capital, manufactures and services (Cozens). Being at the centre between the powerful economies of the Indian subcontinent and the Far East, and within the oil-rich Gulf area, the IO occupies a strategic position to facilitate a tremendous amount of trade within this expanse. This, along with the rim's vast consumer market and trade prospects, makes the IO region an economically vibrant area and the Ocean a vital trade facilitator. The population in the IO rim is close to two billion, providing an enormous market for trade and consumerism. The movement of trade and population in the IO region has combined with geoeconomics to create a common geographical space of tremendous economic prospect and global importance.

The IO features four critically important access waterways facilitating international maritime trade - the Suez Canal in Egypt, Bab-el-Mandeb (bordering Djibouti and Yemen), the Straits of Hormuz (bordering Iran and Oman), and the Straits of Malacca (bordering Indonesia and Malaysia). These "chokepoints" or narrow channels are critical to world oil trade as huge amounts of oil pass through them (Energy Information Administration, 2004). The strategic Persian Gulf is a passageway of significant interest to all those states whose economic and strategic interests require safe trading routes in and out of West Asia. These paths, without exaggerating the fact, not only act as the strategic and economic lifeblood of many states along them but play the role as the main arteries for global trade and energy supply, the majority of which are transported by sea. Hence, the importance of the IO to facilitate the world's trade and its contribution to global economic prosperity cannot be overemphasised.

The trade routes passing through the IO are gaining prominence with increasing dependence by the world's economic powers on energy supplies from the Gulf. It has been estimated that European, Japanese and US economies respectively import 70%, 76% and 25% of their crude oil requirements from the IO rim. China, fast emerging as one of the world's most powerful economies, also has vital trading and energy interests in the IO waters and rim. Adding to this is the substantial economic progress in developing states along the IO such as in India and South Africa. Upward trends in shipping and international trade have seen foreign trade emerging as a matter of national priority to many states, further accentuating the ocean's importance as a global trade facilitator (Pant, 2005).

The economic development of many countries in the IO rim is closely related to seaborne trade. They depend on the free passage of goods across the seas, and the majority of Asia-Pacific countries in the IO region, with their export-oriented economic structure, more than ever depend on maritime transportation (Guoxing, 2000). There has been an unmistakable trend of previously inward-looking countries in the IO region opening their economies at many levels - bilateral, regional and multilateral. This development has given rise to new opportunities for international trade, much of which is transported by sea. With this background, the IO will continue to play a prominent role in facilitating global maritime trade.

Development of Ports and the Shipping Sector in the IO Rim

In mapping the increasing importance of the IO as a global maritime trade enabler, the fact that the bulk of the world's shipping passes through the region has to be emphasised. Increasingly, with trade liberalisation, maritime trade passing through the ocean has dramatically increased as a result of the export-import orientation of many trading countries and their energy imports. The IO provides maritime advantages to several of its littoral states in terms of their strategic location along the Ocean. These points decisively highlight the significance of the IO in facilitating global maritime trade and its importance from a strategic viewpoint.

Amidst increasing transnational commerce, the shipping lanes of the IO have assumed a pivotal significance to facilitate this development. Global trade growth has heightened the importance of this vital ocean, with its strategic waterways and links with major maritime trade routes, in linking the trade of countries and economic regions. Testimony to this importance is the tremendous growth of ports and the shipping sector in the IO region. In sync with the regional and global trade explosion, many maritime infrastructures such as ports and shipping facilities have been built and upgraded along the IO rim to facilitate the trading boom. A glance at the IO map reveals many important harbours and seaports dotted along the region's shores, stretching from Durban in South Africa to Melbourne in Australia. These ports are critical trade facilitators to the economies of the respective countries and to the IO region.

As more IO countries are engaged in international trade with distant neighbours, more of the world's goods will be transported via the ocean. In concurrence with this trend, trade volumes have been increasing steadily in the IO region. Over US$100 billion of China's trade alone passes through the IO annually, and this figure is projected to grow rapidly (IPCS, 2005). Bilateral trade between China and India, two of the world's emerging economic powerhouses, stood at USD13 billion in 2004, representing 1% of China's global trade and 9% of India's (Mitra and Thompson, 2005). There has also been a significant rise in trade volumes moving between post-apartheid South Africa and other IO states such as India, Malaysia, Singapore and Australia, and there is substantial potential for future growth (Ngubane, 1997). Following the far-reaching changes in South Africa's economic, social and political policies, the country is poised to become a powerful engine for growth among countries in sub-Saharan Africa and the island states along the traditional Cape route. India's emergence as an economic power, with a population of a billion, provides another tremendous opportunity for further economic and trade growth in the IO region. It is posited that these developments will have tremendous spillover effects to the growth of the maritime sector across the IO region to support trade expansion. Regional initiatives such as the formation of the Association of Shipping and Port Authorities in the Indian Ocean by the Indian Ocean Marine Affairs Cooperation (IOMAC) will further boost efforts to promote port and shipping development in the IO region (UNU, 2005).

The demand for new ships, as a result of booming global trade and mandatory phasing out of ageing tankers by the International Maritime Organization, has resulted in a boom in new building orders. The collapse of European shipyards and the cost-competitiveness of Asian shipyards have resulted in a dramatic shift of the shipbuilding industry from the west to east. This has had a positive effect in the development of shipping in the IO region, to a certain extent. Although the shipbuilding industry is dominated by yards from the Big Three shipbuilding giants of Asia - China, Japan and South Korea - several IO states have also got into the act. The combined share of the world's orders from yards in India, Pakistan and Bangladesh stood at nearly 4% in 2004 (Exim-India, 2005). India has taken a cue from China to embark on a major shipbuilding program by promoting new shipyards and upgrading existing ones.

It has been argued that increased investment in port facilities and upgrading shipping infrastructure in the IO region would significantly boost regional trade (Puri, 2003, pp. 306). The projected upward trend in cargo traffic worldwide will require massive investments for developing and improving port and shipping facilities in the region to cope with the growth. Trading states in the IO rim need to upgrade their national lines to meet increasing demand for shipping services and to capitalise on container traffic growth. A more modern maritime sector would provide a catalyst to higher trade volume and brighter economic prospects for many IO countries which are economically dependent on the Ocean.

Prospects for Further Growth in Maritime Trade in the IO Region

The IO has played many important roles in the history of the IO region and will indeed continue to chart the destiny of its rim states. However, the sense of esprit de corps amongst the IO littoral and island states is still weak when compared, for example, to the economic and other ties that bind together the countries of the Asia-Pacific. This is perhaps due to the economic and socio-political chasm that separates the states across the expanse of the IO region. However, comfort can be drawn from the fact that the past decade has seen a noteworthy expansion in intra-regional trade and investment among the IO rim countries.

While many IO countries have remained inward-looking, some have moved significantly towards adopting open and outward-looking economic and trade policies, though differing widely in terms of scope, pace and policy instruments. Despite the variability, there are some common features characterising these changes, namely, ensuring macro-economic and fiscal stability through investment-friendly policies, elimination of non-tariff barriers, removal of exchange controls and privatization (Fatemi, 1996). Economic, trade and political trends will continue to change and influence the development of maritime trade around the IO rim and internationally. The dynamics of the world in flux, led by the pervasive effects of liberalisation and globalisation, will continue to affect developments in the IO region and encourage growth in intra-regional maritime trade and investment in the area.

Although it has been mentioned that the IO region has enjoyed remarkable intra-regional trade growth, it is sobering to note that trade volumes in the IO rim are coming off a very low base. As such, the levels of intra-regional trade are not significant relative to trade outside of the region. Perhaps initiatives to develop more intra-IO trade relationships and economic cooperation, and efforts to boost foreign direct investment into needy IO rim economies, will help correct the imbalance. These would achieve the effect of stimulating economic and trade growth in the region, and indirectly boost the development of its maritime sector as one of the pillars of facilitating trade in the region.

It is observed that the maritime sector is undergoing a process of consolidation and concentration evidenced through mergers and alliances among ports and shipping liners, and the construction of increasingly large container ships. All of these developments will demand greater capacity and investment in ports and shipping, and will place a premium on reaching economies of scale. As this trend is foreseen to dictate developments in the maritime sector in the years to come, it will pose keen challenges to the IO region as a crucial facilitator of global maritime trade. IO states will have to adopt international trade and maritime agreements and conventions, attract investments, and adopt new technologies and systems to improve operational capacity, among others, to remain competitive. For IO states to develop competitive maritime sectors, they must strive to decentralise and privatise port services to enhance their commercial competitiveness, enhance infrastructural, supply chain and logistics capacity and capabilities to support growing port throughput, and generally improve in many other areas of operations. With the right kind of policies and foreign investments, it is anticipated that states in the IO rim will see progressive growth in their ports and shipping sectors.

The World Trade Organisation (WTO), whose efforts on trade facilitation resonate loudly in the IO region, has identified a number of key barriers which need to be overcome to promote international trade and commerce. Once these hurdles are surmounted, one can imagine the burst of growth in investment and in trade facilitated by the maritime sector via the IO sea lanes. The ever-changing world's trade structures and flows resulting from WTO initiatives have given rise to concomitant effects on global port development and shipping patterns, leaving their mark in the IO and along its rim. While developed countries in Europe and North America are still fuelling the world's merchandise trades, the balance of power has significantly shifted to the Asian region, namely to China and India. These two states, along with so many others, depend largely on the IO to facilitate their trade and energy supply to fuel their economic growth.

It is hoped that stronger emphasis on trading by IO states could lead to the opening up of new trade opportunities and the growth of greater economic prosperity in the IO region. Given the evidence in the analysis of the factors behind the IO's influential role in spurring global maritime trade, these objectives are well within reach. Foreseeable future trends and patterns in maritime, economy, investment, capacity building and trade give ample ground to be optimistic of the IO's continued importance in facilitating world seaborne trade and as a strategic waterway to many states. They also point to greater market-driven growth in trade, investment and economic development in the IO rim, befitting the ocean's strategic location, importance and influence in promoting global maritime trade.

Conclusion

As the importance of maritime trade and strategic interests increase, the IO will continue to be at the forefront of attention of trading states and the world's great powers. This will exert plenty of challenges to this Ocean in areas such as navigational safety, environmental integrity, sovereignty and security, and enhance the focus on the impacts they will have on society, trade and economic development in the region. Barring drastic developments, the near future will undoubtedly see an increase in maritime activities in IO waters, fuelled by factors such as increasing international trade among states, demographic changes, integration of economies, modernisation in ports and shipping, increasing dependence of economic powers on energy supply, and growing strategic interests. Subsequently, the scenario in the IO will become more complex and the challenges greater than ever.

It is expected that the sea lanes of the IO will become even busier in the future as global trade grows in size and importance. Maritime transport, as the most economical and effective means of transport to support international trade, will correspondingly see an increase in activities, and the IO, as a crucial trade waterway, will in tandem grow in prominence and significance. Increasing regional cooperation such as those initiated on the platform of IOR-ARC will enable IO countries to participate more actively and effectively in the global economy and leverage their membership in the grouping. These developments and other trends described in this chapter indicate greater demands and bigger roles being placed on the IO to facilitate global maritime trade in the future.

The new world order, under the dictates of globalisation and trade liberalisation, will necessitate IO countries to unite on a single platform to promote their economic and trade interests. The strategic importance of the Ocean and its region is set to heighten as global economic integration gains momentum (Taib, 2003). Its pivotal role will ensure that it will remain in the sharp focus on the radar of intra and extra-regional powers (Kohli, 1996).

An expanse as great as the IO that divides continents could also prove to be a unifying force of people, given their commonalities - united under an oceanic community of states. Perhaps the people in the IO rim could seek inspiration in the old quote, *"the land divides, but the sea unites"*.

References

CIA web site (2005), http://www.cia.gov/cia/publications/factbook/geos/xo.html

Cozens, P. 'Maritime Security in the Indian Ocean' by Peter Cozens, Fellow at Asia Pacific Research Institute. Overview prepared for a research project planned in cooperation with the South Asia Institute of the University of Heidelberg, Germany.

Curtin University IOC web site (2005), http://www.curtin.edu.au/curtin/centre/ioc/ioripage.htm

Energy Information Administration (2004), World Oil Transit Chokepoints'. Report by Energy Information Administration, Washington DC, March.

Exim-India (2005), From http://www.exim-india.com/link/htmls/shipping.htm

Fatemi (1996), 'The Indian Ocean Region: An Attempt at Economic Integration'. Draft of a collaborative research project prepared for the International Conference on Economic Integration in Transition, Athens, Greece, 21-24 August.

Guoxing, J. (2000), 'SLOC Security in the Asia Pacific', *Center Occasional Paper*, Asia-Pacific Center for Security Studies, Honolulu, Hawaii.

IFIOR (1995), 'Current Economic Characteristics and Economic Linkages in the Indian Ocean'. Working paper presented by Australia's Department of Foreign Affairs and Trade at the International Forum on the Indian Ocean Region. 11-13 June, Perth, Australia.

IPCS (2005), http://www.ipcs.org/printSeminar.jsp?action=showView&kValue=1602

Kamat (1998), 'The Relevance of the Indian Ocean', *The Daily*, 5 April.

Kohli (1996), 'Maritime Power in Peace and War - An Indian View', *African Security Review*, Vol. 5 (2).

Mitra and Thompson (2005), 'China and India: Rivals or Partners?', *Far Eastern Economic Review*, April.

Ngubane, B. (1997), 'RIM Relationships and Economic Development', speech at the Indian Ocean Rim Conference, Durban, South Africa, 10 March.

Pant, N. K. (2005), Indian Ocean: a strategic appraisal', sighted at http://www.geocities.com/siafdu/20.html

Puri (2003), 'An Ocean of Opportunities', *In Regional Cooperation in Indian Ocean : Trends and Perspectives*, Edited by P V Rao, South Asian Publishers: New Delhi.

Rumley, D. and Chaturvedi, S. (2004), 'Changing geopolitical orientations, regional cooperation and security concerns in the Indian Ocean', in Rumley and Chaturvedi, eds., *Geopolitical Orientations, Regionalism and Security in the Indian Ocean*, New Delhi: South Asian Publishers, pp. 21-31.

Sakhuja, V. (2003), 'Indian Ocean and the Safety of Sea Lines of Communications', Institute for Defense Studies and Analyses, India, 18 July.

Taib, M. (2003), 'Indian Ocean Region: Malaysia's Perspective'. Paper presented at the Indian Ocean Conference, Hawaii, 19-21 August.

UNEP (1999), 'Western Indian Ocean Economic Outlook'

UNU (2005), From http://www.unu.edu/unupress/unupbooks/uu15oe/uu15oe0n.htm

Wikipedia web site (2005), http://en.wikipedia.org/wiki/Indian_Ocean

Chapter 7

The Uranium Trade in the Indian Ocean Region

Dennis Rumley and Timothy Doyle

Introduction

It is an interesting irony that, on the one hand, apart from South West Asia and South Asia, the Indian Ocean is surrounded by nuclear weapons free zones (Antarctic Treaty, Treaty of Bangkok, Treaty of Pelindaha, Treaty of Rarotonga) while, on the other hand, it is fast becoming a nuclear Ocean (Doyle, 2005). Apart from the increasing number of regional nuclear weapons on land, as well as the indeterminate number on and under the Ocean itself at any one time, the increasing global and regional demand for nuclear energy is having a significant impact on the structure of the Indian Ocean uranium trade (Table 1). These impacts, in turn, raise a host of security questions linked to nuclear safety, uranium flows, the flows and storage of nuclear waste and the security of sea lanes of communication (SLOCS) in the Indian Ocean Region.

Table 1
Indian Ocean Region: Nuclear Energy Users, Uranium Suppliers and Nuclear Waste

Nuclear Energy Users

Indian Ocean Region	Actual - India, Pakistan, South Africa
	Potential - Indonesia, Iran
Extra-Regional Impact	China, France, Japan, South Korea, UK, USA

Uranium Suppliers

Indian Ocean Region	Australia, India, Iran, South Africa
Extra-Regional Impact	Namibia

Nuclear Waste

Indian Ocean Region	'Dumping' nuclear waste - Africa
	New Users
	Potential Regional Depositories
Extra-Regional Impact	Japan-France

Global and regional nuclear energy demand and supply

In 2002, the energy mix of Indian Ocean states indicated the minimal regional use of nuclear energy. Indeed, most IOR-ARC member states did not produce any nuclear energy, while South Africa and India produced relatively small amounts (Table 2). By 2005, however, South Africa produced 6.1% of its electricity requirements from nuclear energy, compared with 3.3% for India and 2.4% for Pakistan (World Nuclear Association).

Table 2
IOR-ARC Energy Mix 2002
(% primary energy consumption)

	Oil	Natural Gas	Coal	Nuclear	Hydro
Australia	33.7	19.1	43.8	0	3.4
Bangladesh	24.1	71.6	2.8	0	1.4
India	30.1	7.8	55.6	1.4	5.2
Indonesia	50.0	30.6	17.4	0	2.1
Iran	45.8	52.6	0.7	0	0.9
Malaysia	43.4	46.9	6.4	0	3.3
Singapore	95.7	4.3	0	0	0
South Africa	21.6	0	74.9	2.7	0.8
Thailand	51.2	33.8	12.5	0	2.3
UAE	25.9	74.1	0	0	0

Source: Rumley, 2005

This relatively rapid increase in the importance of nuclear energy for regional states is part of a broader global trend implicating at least three sometimes overlapping groups of states - those which are energy dependent, those which are industrialising and those that are interested in maintaining or developing their own nuclear technology. In all three groups, however, the development of nuclear energy has important implications for the spread of nuclear weapons, since, although the nuclear Non-Proliferation Treaty (NPT) is at pains to emphasise the legitimacy of the development, use and application of nuclear energy "for peaceful purposes", there is now "a very fine line between the civilian and military applications of the nuclear fuel cycle" (O'Neil, 2005). It would therefore be remiss not to make the point that, in all three groups of states, there is an increasing link between nuclear energy and the militarisation of the Indian Ocean Region.

Projected Global Nuclear Energy Consumption

The United States Energy Information Administration (USEIA) provides projections for nuclear energy consumption to 2025 and these have been extrapolated for the world's largest oil consumers (Table 3). While many industrialised European states currently do not plan significant nuclear energy expansion (for example, Belgium, Netherlands, Spain, Sweden, Switzerland and the UK), two of these states - Germany and Italy - are projected to consume no nuclear energy by 2025. In addition, nuclear consumption is projected to decline in Canada and Russia, while in the remaining six states, nuclear energy consumption is projected to increase. This is especially the case in absolute terms for France (+103 bn kwh), China (+88 bn kwh) and South Korea (+79 bn kwh). However, the largest percentage increases will occur in China (+133%), South Korea (+56%) and India (+43%). France, India and South Korea have all been described as "energy import dependent states" (Rumley, 2005).

Table 3
Largest Oil Consumers: Nuclear Energy Consumption 1990-2025
(billion kilowatthours)

	1990	2010	2025
Canada	69	108	98
China	0	66	154
France	298	447	550
Germany	145	137	0
India	6	46	66
Italy	0	0	0
Japan	192	369	411
Russia	115	141	99
South Korea	50	141	220
USA	577	794	812

Source: International Energy Outlook 2004, US Energy Information Administration

As is well known, combating human-induced climate change requires states to reduce the carbon intensity of their economies, and it has been argued that, by 2025, G8 members should generate at least 25 per cent of all of their electricity from renewable sources (ICCT, 2005). However, to some commentators, climate change is happening so rapidly that a 2025 deadline will be too late to avoid its most serious implications, and that, in fact, the only major practical solution to solving global warming is for the rapid expansion of nuclear power (Rumley, 2005). This argument is especially relevant for the ten states possessing the world's largest absolute volumes of carbon dioxide emissions - USA, China, Russia, Japan, India, Germany, Canada, South Korea, Italy and France. As we have seen, of these states, only China, India and South Korea were projected to have significant increases in nuclear power over the next two decades (Table 3). However, the 2004-5 'oil shock' has prompted the United States to seemingly reassess its position on nuclear energy, and, in a speech in April 2005, the US President declared that, "a secure energy future for America must include more nuclear power" (Rumley, 2005). Whereas the United States had not planned any increase in the number of nuclear reactors, this speech implies a likely increase from its present complement of 103 (World Nuclear Association).

This argument, which is heavily endorsed by the public relations position of the nuclear industry itself, is enjoying wide currency amongst policy-makers in the Indian Ocean Region and beyond. Environmental Non-Governmental Organisations (ENGOs), on the other hand, are also spending an inordinate amount of time and resources trying to refute the uranium and nuclear energy industry's 'greenhouse friendly' status. In a paper released simultaneously by Friends of the Earth, the Australian Conservation Foundation and Greenpeace in 2005, the industry argument is refuted on the basis that it is widely accepted that emissions reductions of the order of 60% are required by 2050 in order to stabilise concentrations of greenhouse gases to avert significant climate change. The July 2005 draft report reads:

> Because of economic and public accountability problems, and nuclear power's limited potential other than in electricity generation, the potential for nuclear power to contribute to reducing greenhouse emission is limited. A doubling of nuclear power by 2050 would reduce greenhouse emissions by about 5% - less than one tenth of the reductions required to stabilize atmospheric concentrations of greenhouse gases (FoE, ACF, GP 2005, unpub.)

Whatever the scientific case, the public relations battle between nuclear industry advocates and environmental NGOs is heating up, and will be a common argument occurring within media outlets in the region over the coming decade, ultimately shaping national and regional public policy.

For extra-regional states, however, adopting a 'nuclear view' of climate change and its potential impact on the energy mix, coupled with the current structure of the Indian Ocean nuclear energy trade imply that, in future decades, the Ocean will be increasingly used as a 'uranium highway' for Europe (primarily France and the UK) and more especially for Northeast Asia (China, Japan and South Korea). This latter point is emphasised in the large number of nuclear reactors (44) which are planned or proposed for Northeast Asia, which represents 29% of the global total (Table 4).

Table 4
Extra-Regional Nuclear Reactors 2006

	Operable	Under Construction	Planned/Proposed
China	10	5	24
France	59	0	1
Japan	55	1	12
South Korea	20	0	8
UK	23	0	0
USA	103	1	13
Global Total	441	27	153

Source: World Nuclear Association

Estimated uranium demand from these five states in 2005 is 26,342 tonnes of uranium, which is 39% of global requirements and 74% of global uranium production in 2003 (World Nuclear Association).

Potential Indian Ocean Nuclear Energy Consumption
Apart from the current small amounts of nuclear power generated by India and South Africa, one possible regional scenario from this perspective of climate change is that several other Indian Ocean states will also seriously consider the development of nuclear power programmes. In any event, both India and South Africa plan to substantially increase their number of nuclear power stations and Iran has already indicated its intention in this regard (Table 5).

Table 5
Indian Ocean Region Nuclear Reactors 2006

	Operable	Under Construction	Planned/Proposed
India	15	8	24
Indonesia	0	0	4
Iran	0	1	5
Pakistan	2	1	2
South Africa	2	0	25

Furthermore, Indonesia, even though in favour of increased regional cooperation in the field of renewable energy, continues to debate the development of nuclear power, an option that it shelved in 1997 (Rumley, 2005). As it stands, the current plan is for Indonesia to build at least four nuclear reactors and to begin the construction of its first nuclear power station in 2010, with a view to commencing nuclear power production in 2016. In addition, Pakistan, which now produces a small amount of nuclear energy, plans to increase its number of nuclear reactors to five (Table 5). The number of planned or proposed reactors for the Indian Ocean Region (60 or 39% of the global total) exceeds that for all of Northeast Asia (Tables 4 and 5). It may well be that Australia will also pursue the nuclear energy option in the future since Prime Minister Howard believes this to be "inevitable".

Uranium suppliers
The current largest uranium suppliers are not necessarily those with the largest uranium reserves (Table 6). For example, Canada, with the third largest global reserves (14.1%) was the world's largest uranium producer in 2002 with 29.2% (Uranium Information Centre). On the other hand, the two states with the world's largest reserves - Australia and Kazakhstan - together produced just over 30% of the world's uranium in 2002.

Table 6
Global Uranium Reserves and Production 2002
(% of global total)

Uranium Reserves	Uranium Production
Australia (27.8%)	Canada (29.2%)
Kazakhstan (15.2%)	Australia (21.2%)
Canada (14.1%)	Kazakhstan (9.2%)
S Africa (9.6%)	Niger (8.8%)
Namibia (7.6%)	Russia (8.8%)
Brazil (6.3%)	Namibia (5.7%)
Russia (4.2%)	Uzbekistan (4.9%)

Sources: Rumley 2005; Uranium Information Centre

In 2002, Australia exported virtually all of its 7637 tonnes of uranium oxide to the northern hemisphere, with only five states receiving in excess of 500 tonnes - USA (2790), Japan (2688), South Korea (756), UK (576) and France (503) (ERA web site). India, on the other hand, while possessing about 2% of world uranium reserves, has the second largest global reserves of thorium, and produces almost entirely for the domestic market. One of the long-term goals of its nuclear programme is to develop an advanced heavy-water thorium cycle (World Nuclear Association).

From a global perspective, both Australia (with 27.8% of reserves and 21.2% of production) and South Africa (with 9.6% of reserves and 2.1% of production) could be regarded as 'underproducers', and, given their relative political stability, are both likely to be future sources for growing Indian Ocean and Northeast Asian demand. Consequently, political pressure from various quarters is mounting for Australia to change its current uranium mining policy. While the current Federal Coalition government seemingly favours a review and a wide ranging policy debate, the opposition Labor Party in the past has preferred uranium mining to be restricted to three mines. However, there is no complete unanimity across party lines on this issue. Nonetheless, the consensus at the Federal level in Australia is moving towards an expansion of uranium mining and export. Indeed, such an expansion is regarded in some quarters as one step towards Australia becoming an "energy superpower" (Shanahan, 2006).

Constitutionally, the Australian States currently possess the power over the extraction of minerals resources (so-called "minerals power"), and uranium mining is presently disallowed in all States and Territories except South Australia (which contains Australia's largest deposits at Olympic Dam) and the Northern Territory. This issue is rapidly developing into an important Federal-State conflict in Australia as States such as Western Australia, with significant uranium deposits and no uranium mine, will likely become increasingly at odds with a Federal government keen to open up new mines and to expand Australia's uranium exports. Political pressure will also be backed up by powerful economic pressure, since, in WA, BHP-Billiton holds two of the largest uranium deposits at Yeelirie and Kintyre.

However, the debate in Australia extends beyond uranium mining into the 'value added' process of uranium enrichment and also into the prospect of nuclear power. The Federal Government currently awaits an energy report due in November 2006 on which a decision will likely be based to go further than uranium mining. It has been suggested that this report will provide the basis upon which the Government will decide in favour of uranium enrichment as well as nuclear power for Australia. However, BHP-Billiton has already cast serious doubt on the economic viability of uranium enrichment in Australia. Rio Tinto has also indicated that currently there is "no supply deficit in the enrichment market" (Johnson, 2006).

Transboundary movements of nuclear waste: the legal basis

While controversy over uranium mining, enrichment and nuclear power is likely to continue to fuel debate in Australia and elsewhere in the region over the next several years, it is the question of nuclear waste which is arguably the most socially, economically and environmentally contentious. Part of the problem in this debate is that no universal definition in fact exists of "waste" and national definitions vary considerably and are often subjectively based. One *legal* definition is that waste comprises "substances or materials which are disposed of, or are intended to be disposed of, or are required to be disposed of by the provisions of national law". "Hazardous waste", on the other hand, consists of materials of varying categories and having specific characteristics that are generally considered harmful to human health (Bamako Convention, article I, annex 1 and annex II).

For the most part, the transboundary movement of hazardous waste has followed the "path of least resistance" principle. Furthermore, there is generally an"economic and regulatory imbalance" between the generating and the importing states. Thus, dumping at sea is easier and less costly, and the developing world, particularly Africa, has long been seen as an attractive recipient region on account of relatively inexpensive disposal costs (Krummer, 1999, 4-7; Lipman, 2002). However, post-Cold War international law, especially in the form of the Basel Convention, has been developed in order to regulate the transboundary movement of hazardous waste.

The Basel Hazardous Waste Convention
The Basel Convention on the Control of Transboundary Movements of Hazardous Wastes and their Disposal came into force in 1992 and, at April 2005, had been ratified by nearly all Indian Ocean states. Apart from the United States, notable regional exceptions included Afghanistan, Bhutan, Burma, Iraq, Laos, Somalia, Sudan Swaziland and Zimbabwe.

One of the more important concerns lying behind the Convention was the fact that, of the more than 400 million tons of hazardous waste produced each year, approximately 80% came from OECD countries and a large proportion of that was being 'dumped' into industrializing states (Greenpeace, 1998). The scope of the Basel Convention, however, does not extend to the transboundary movement of radioactive materials although it does allow some transboundary movement of hazardous waste "only when the transport and the ultimate disposal of such wastes is environmentally sound". This clause has been used as a bargaining position for those states possessing hazardous waste exports since it could imply unlimited external access so long as appropriate local environmental expertise and technology existed. An important amendment to the Basel Convention, however, adopted in 1995, prohibited the export of all hazardous wastes from richer to poorer states from 1998.

The issue of the transboundary movement of nuclear waste was explicitly not inserted within the Basel Convention because its movement was seen as falling within the sphere of competence of the International Atomic Energy Agency (IAEA). The adoption in 1990 by the General Conference of the IAEA of a non-binding code of practice on the International Transboundary Movement of Radioactive Waste, which was proposed by UNEP, however, has in fact resulted in a weakening of the principles underpinning the Basel Convention (Krummer, 1999, 85).

The Bamako Convention
While Africa possessed the dubious distinction of being "first choice" for the dumping of European nuclear waste and persistent organic pollutants (POPs) in the 1980s, it was the first to respond politically to the threat of "waste colonialism" (Bernstorff and Stairs, 2001, 4). Prior to the ratification of the Basel Convention, many African states were especially concerned about the transboundary movement of such hazardous waste into Africa from industrialised countries and some indeed saw this process as one of the systematic 'dumping' of nuclear waste into Africa (Krummer, 1999, 99). At the May 1988 Organisation of African Unity (OAU) Council of Ministers 48th Ordinary Session in Ethiopia, a resolution condemned the importation into Africa of industrial and nuclear waste as a "crime against Africa and the African people" and called upon member states to introduce import bans. The resolution condemned "all transnational corporations and enterprises involved in the introduction, in any form, of nuclear and industrial wastes in Africa; and DEMANDS that they clean up the areas that have already been contaminated by them" (Organization of African Unity Secretariat, CM/Res.1147-1176, 1988).

As a consequence of this resolution, work began on an African Convention under the auspices of the OAU shortly after the adoption of the Basel Convention, since the latter excluded nuclear waste. There was therefore a concern that certain of the needs of African states were not properly taken into account, and thus, while the Basel Convention was a Convention of the North, there was need for a Convention of the South. The resultant Bamako Convention, which was adopted in Mali in January 1991, entered into force in April 1998.

Clearly, adherence and compliance are major problems, since, of the 53 African Union states, only 28 had signed and only 21 had ratified the Convention at 1 April 2005. Furthermore, the Convention has been ratified by only 8 of a potential 21 African Union Indian Ocean members (Table 7).

Table 7
Status of the Bamako Convention: African Union Indian Ocean States

	Signed	Ratified
Botswana	-	-
Burundi	1991	-
Comoros	2004	2004
Djibouti	1991	-
Ethiopia	-	2003
Kenya	2003	-
Lesotho	1991	-
Madagascar	2004	-
Malawi	-	-
Mauritius	-	1992
Mozambique	-	1999
Rwanda	1991	-
Seychelles	-	-
Somalia	1991	-
South Africa	-	-
Sudan	-	1993
Swaziland	1992	-
Tanzania	1991	1993
Uganda	-	1998
Zambia	-	-
Zimbabwe	-	1992

Source: African Union web site

Of particular international concern are the five states that have neither signed nor ratified the Convention, as well as a further seven, which have signed but are yet to ratify. It may be that some states have stalled either signing or ratification in order to participate in the lucrative trade in hazardous waste (UNEP, 2000). This may well be true for the six Indian Ocean littoral states of Djibouti, Kenya, Madagascar, Seychelles, Somalia and South Africa, none of which were in the original Convention signatory group of twelve states. Furthermore, of these six littoral states, two have neither signed nor ratified Bamako (Seychelles and South Africa) and a further two (Djibouti and Somalia) have yet to ratify either the Basel or Bamako Conventions.

Dumping nuclear waste: the example of Somalia
The Indian Ocean tsunami of December 2004 resulted in the washing up on

Somalian beaches of many containers of nuclear and toxic waste that were illegally dumped during the early 1990s (Clayton, 2005). It is alleged that, in at least one case, a lucrative financial agreement had been reached between the interim government headed by Ali Mahdi Muhammed and certain Swiss and Italian companies to import into Somalia from Italy millions of tonnes of nuclear waste (see Table 3). These companies were alleged to be under the control of the Italian mafia, and the Somalian deal was said to be only one part of so-called "eco-mafia" operations (Grosse-Kettler, 2004, 29). For the Europeans, the cost per tonne (US$8) represented a fraction of the likely cost of up to US$1,000 per tonne of appropriate local treatment and disposal (Clayton, 2005).

Given Somalia's strategic location and its current statelessness, such illegal or unauthorised movements of nuclear waste have potentially very significant implications not only for the human security of the Somalian population, but for the Indian Ocean environment, the Indian Ocean routes along which such flows take place, as well as the lethal prospects of the potential terrorist use of such nuclear materials. It has been noted that, in fact, Somalia is a "stateless war economy", one of the requirements of which is to engage in international "commercial complicity" since its local economy is unable to meet military expenditures. Funding the war economy is achieved in various ways, including via trade by local conflict groups with international corporations and institutions in unauthorised commodities, including nuclear waste. Indeed, Somalia currently functions as a transhipment point and a supply route for a wide variety of illegal merchandise for the whole of the Horn of Africa and beyond (Grosse-Kettler, 2004).

This touches on another fundamentally important security issue for the Indian Ocean Region. It has recently been pointed out that seizures of smuggled radioactive material capable of making a terrorist "dirty bomb" have doubled in the past four years. According to the IAEA, smugglers, mainly from the former Eastern bloc, have been caught attempting to traffick such materials on more than 300 occasions since 2002, most of the incidents are understood to have taken place in Europe (Smith, 2006). IOR-ARC potentially has a very important role as a pressure group, both regionally and in international forums, to try and eradicate the smuggling of radioactive materials into the Region and to prevent its dumping in the Ocean.

The Movement of Nuclear Waste Through the Indian Ocean

At the regional level, it is not just waste that is generated within the region, but also the ever-increasing volume of waste that passes through the region via sea routes.

As a consequence, the Indian Ocean Region is currently being used as a key transport route for nuclear waste. Much of this transport revolves around the servicing of the two biggest destinations for the processing of waste: Sellarfield, in England and Le Hague in France. Most of this waste in the region is produced by Japan. Song writes:

The first question in nuclear waste disposal is reprocessing, which can recover a significant quantity of useable material for nuclear reactors and reduce the amount of high-level waste that must be stored to about three per cent of the original spent fuel (Song, 2003: 8).

Waste reprocessing breaks down spent fuel, chemically dissolving it, with the plutonium separated from it. Apart from large, routine discharges of radioactivity from this high level and long-lived nuclear waste, this plutonium is then mixed with uranium (MOX) and re-used in conventional reactors. The Japanese are the biggest proponents of this process, and receive their shipments from both the British and French reprocessing plants. This MOX fuel is transported through the Indian Ocean on its way back to Japan, though few reliable records are available in the public domain documenting this traffic. The few attempts at monitoring these movements have been embarked upon by international environmental non-government organisations (ENGOs) like Greenpeace, in the absence of efforts made at both the nation-state and international levels. It is impossible, therefore, to know just how many of these shipments are being made, due to the high levels of secrecy that surround them (Doyle, 2005). According to Greenpeace, on 21 July 1999, two ships carrying weapons-usable plutonium, left Europe via the Cape of God Hope, through the Indian Ocean to Japan. Greenpeace reported as follows:

> The two British flagged vessels, the Pacific Teal and the Pacific Pintail, left Barrow in Britain and Cherbourg in France carrying the first commercial shipment to Japan of mixed-oxide (MOX) reactor fuel, made from plutonium and uranium. An estimated 446 kilograms of plutonium is contained in the 40 nuclear fuel elements - enough fissile material to construct at least 60 nuclear bombs... The Cape of Good Hope has become the path of least resistance (Greenpeace, 1999).

Only Mauritius, acting alone, made public its opposition to the reprocessed fuel's transport, by refusing to admit the vessels into its Exclusive Economic Zone (African News Agency 7/3/99). Of course, this can only be a symbolic position, as the devastation which would occur if the vessel were sunk, or caught on fire, would make the 200 km zone, as a zone of protection, look ludicrous. The point has already been made that under existing liability agreements, there is some limited compensation under international conventions; "but no assurances exist whatsoever that the full costs of health, environmental and economic damages would be paid to victims in en route states" (Greenpeace, 1999).

Since the turn of the millennium, limited surveillance by ENGOs of Indian and Pacific oceanic nuclear flows has continued in a piecemeal fashion, with Greenpeace Australia concentrating on shipping movements from Albany, Sydney and Tasmania. However, even this limited surveillance provided by ENGOs has now declined. According to the Greenpeace Australia's current nuclear campaign co-ordinator - James Courtney - the costs associated with aerial surveillance have proved increasingly prohibitive for the not-for-profit organisation (Courtney, pers. comm., 2005).

Another key event that needs to be mentioned here, which may dramatically effect the flows of nuclear waste in the region is the decision by Japan to commission its own reprocessing plant. As the data show earlier in this chapter (Table 4), Japan has now embarked upon a new 10-year energy plan, calling for the expansion of national nuclear energy by approximately 30% (EIA 2004, online). Not only will this include the construction of up to twelve new nuclear plants on top of its 54 existing reactors, but the construction of a reprocessing unit which will convert spent fuel into mixed oxide plutonium and which will be reused, in some cases, in specifically designed reactors. At this stage, it is envisaged that the reprocessing plant will only be used to treat waste produced from the Japanese domestic market. If this is the case, it is likely that shipments to Sellarfield and La Hague may decline in number. On the other hand, if the reprocessing unit opens up its facility to the international market (particularly to the emerging facilities in less affluent Indian Ocean states) - converting waste sourced from non-Japanese facilities - this may actually substantially increase waste and reprocessed waste flows within the Indian Ocean Region. The ultimate disposal and storage of such waste, however, is an especially problematical regional/global issue.

Nuclear Waste Storage in the Indian Ocean Region

At this juncture, the environmental security concerns of waste storage and transport become intermeshed, once more, with more traditional security concerns of nation-states and their defence. Since spent nuclear fuel can be reprocessed to produce both weapons of mass destruction - as well as energy - the placement of such waste dumps (as well as reprocessing plants) is highly politicised internationally and regionally. For example, the United States is increasingly involved in decisions as to who can and who cannot store and reprocess spent nuclear fuel. In fact, recently, the US decided to store all its own spent fuel at Yucca Mountain by 2010 in

Nevada, in large part due to its fears that exported waste may find its way into the nuclear defence programs of other states (Doyle, 2005).

Australia

Given the Basel Amendment and given Australia's apparent reluctance at the time to support it, trade in hazardous wastes to and from Australia with other OECD states is likely to be a real possibility. A well-publicised report, which surfaced first in December 1998, showed that the Swiss-based multinational mining company, Pangea Resources, following a global site survey and on the basis of a "high isolation" concept, had already targeted inland Australia for the global dumping of nuclear waste. Most of the funding for this endeavour had been provided by BNFL. However, the Australian Industry, Science and Resources Minister at the time stated in Federal Parliament that no Ministerial discussions had taken place on this issue. The WA State Deputy Premier, however, admitted that discussions had taken place with Pangea "about two years ago" and that Pangea believed that the "Centralian Superbasin" in Western Australia was "the perfect storage facility" on account of its "stable political scene and geography".

It was argued that since this sedimentary formation had been stable for 500 million years that it could potentially take nuclear waste from the more than 400 nuclear power stations around the world. It was also suggested that, to a significant degree, the global nuclear waste problem would therefore be solved since this facility would receive something in excess of 75,000 metric tonnes of waste over a forty-year period. Furthermore, storage costs and infrastructure development would generate an enormous economic dividend to Australia of the order of about 1% of GNP per annum. The ensuing public and political opposition to this proposal and the subsequent closure of Pangea's Australian office led many commentators to feel that, by 2002, that the Pangea concept was a non-starter (IAEA, 2004, pp.13). In fact, "Pangea ceased operations in 2001 when the owners decided that the commercial prospects for an international repository were too far into the future to justify the investment required" (ARIUS web site).

During the last two years, differing signals regarding the development of nuclear power and the creation of nuclear storage facilities in Australia have been emerging from Australian politicians. For example, in July 2004, the Australian Prime Minister announced that the Australian government was examining options for the siting of a facility for nuclear waste "with a preference for an offshore site". Furthermore, in September 2004, the Australian Environment Minister, Western Australian Senator, Ian Campbell, stated that the only locations being considered by the Australian government for nuclear waste storage were on "offshore islands".

However, on 6 June 2005, the Minister did not rule out Western Australia as a possibility, since NSW, South Australia and the Northern Territory "had been given assurances already that nuclear waste would not be buried there" (Dortch and Mason, 2005). However, while the NSW Premier appears to favour nuclear power, the NSW government continues to consider the findings of its Joint Select Committee on the Transportation and Storage of Nuclear Waste which reported in February 2004. On the other hand, Western Australia (2004), together with South Australia (2003) and the Northern Territory (2004), have all legislated against the construction and operation of any facility designed for nuclear waste storage. The Queensland State Government, however, appears to prefer to seek clean coal technology for its 300 years supply of coal.

Nonetheless, Pangea Resources, which reconstituted itself in Switzerland in February 2002 into the Association for Regional and International Underground Storage (ARIUS), still believes that Western Australia is "the best location in the world for nuclear waste storage" (Law, 2005). In addition, the 'logic' of this waste debate (which is apparently supported by the United States - The West Australian, 16 May 2006) further rests heavily on a 'moral imperative' - that is, if uranium mines are opened in Western Australia and if large quantities of uranium are exported to Northeast Asia and elsewhere, then WA has a 'special responsibility' to receive the nuclear waste, which is an outcome of its export activity. Some politicians and commentators, however, dispense completely with any moral responsibility argument and accept the economic rationale for Australia both mining uranium and taking the nuclear waste. Indeed, the Federal member for Kalgoorlie is reported to have said that Australia should consider storing high level radioactive waste and that it would be able to "charge like a wounded bull" for this service (ABC News Outline, 2006).

Clearly, such a global respository, if it came into being, would have very significant long-term implications for the environmental security of the Indian Ocean itself and for the security of sea lanes terminating at their Western Australian destination. In the 2005 budget, the Australian Federal Government allocated around A$140 million to protect offshore oil and gas platforms in the North-West Shelf against a potential medium-level terrorist threat. This would allow for two extra patrol boats and trials of non-personnel flights. However, a much more comprehensive security strategy would be needed if a global nuclear waste repository were located in WA.

Other Indian Ocean states
Two other Indian Ocean countries - India and Pakistan - store their own waste. In

the increasingly controversial case of Iran, there appears to be some tension as to whether the waste from the newly-constructed Bushehr nuclear power plant will be transported to Russia, or will remain on home soil (Kerr, 2002, pp. 29). Sections of the Iranian government want to store their own waste (to potentially utilise the enriched uranium as part of a possible weapons program), whilst powerful elements in the international community - particularly the United States - have made it clear that if the enriched waste is not returned to Russia, then the nuclear program within Iran should be terminated.

In South Africa, high-level waste from the country's two nuclear reactors at Koeberg (which generates about 95% of the waste) and Pelindaba is stored on site because it is considered too dangerous to move. Critics of South Africa's nuclear waste policy argue that there is "no real plan" for its disposal despite the release for a brief month-long public comment in 2003 of its radioactive waste-management strategy (Roelf, 2005). Apparently, the policy document identifies three main options for South Africa's waste - above ground interim storage, deep geological disposal and reprocessing. Although deep storage was favoured, there were problems locating a suitable site. Reprocessing, on the other hand, would require shipments to the UK, Japan or France. South Africa has stated that it would never consider accepting nuclear waste for storage from another state, however.

Conclusion

The Indian Ocean is fast becoming the nuclear ocean. Because of its status as the ocean of the south, the environmental security implications of expansion of uranium mining and nuclear power have not been sufficiently explored. The possibilities of nuclear accidents, whether in the mining of uranium, power generation, reprocessing, transport or storage are seen as risks which those less affluent and/or more remote are expected to take. With its increased nuclear profile, it is critical that environmental security concerns are included alongside the more traditional concerns of weapons proliferation and possible nuclear war.

References

ABC News Online (2006), ' WA dump', 24 May, http://www.abc.net.au/news

African Union web site: http://www.africa-union.org/

Albright, D. and Hibbs, M. (1993), 'South Africa: the ANC and the atom bomb', *Bulletin of the Atomic Scientists,* Vol 49 (3), pp. 32-37.

Association for Regional and International Underground Storage (ARIUS) web site, at: http://www.arius-world.org/

Bernstoff, A. and Stairs, K. (2001), POPs in *Africa: Hazardous Waste Trade 1980-2000,* Amsterdam: Greenpeace International.

Clayton, J. (2005), 'Somalia's secret dumps of toxic waste washed ashore by tsunami', http://www.timesonline.co.uk/

De Villiers, J. W., Jardine, R. and Reiss, M. (1993), 'Why South Africa gave up the bomb', *Foreign Affairs, Vol 72 (5), Nov/Dec, pp. 98-109.*

Dortch, E. and Mason, G. (2005), 'WA nuclear waste dump fear revived', *The West Australian* newspaper, 7 June, page 1.

Doyle, T. (2005), 'The Indian Ocean as the *Nuclear Ocean:* environmental security dimensions of nuclear power', in Rumley, D. and Chaturvedi, S., eds., *Energy Security and the Indian Ocean Region,* Delhi: South Asian Publishers, pp. 230-252.

Grosse-Kettler, S. (2004), 'External actors in stateless Somalia: a war economy and its promoters', Paper 39, Bonn International Center for Conversion (BICC).

IAEA (2004), *Developing Multinational Radioactive Waste Repositories: Infrastructural Framework and Scenarios of Cooperation,* IAEA-TECDOC-1413, Vienna.

Jognson, C. (2006), 'Inquiry set to back uranium enrichment', *The West Australian* newspaper, 6 October, page 6.

Krummer, K. (1999), *International Management of Hazardous Wastes,* OUP.

Law, P. (2005), 'New bid for WA N-dump', *Sunday Times newspaper,* 12 June.

Lipman, Z. (2002), 'A dirty dilemma: the hazardous waste trade', *Harvard International Review,* Winter 2002. Vol. 23 (4), pp. 67-71.

Michelmore, A. (2004), 'Olympic Dam's Position in the World Uranium Industry', address to the Citigroup Global Markets, uranium information session, Sydney, 14 December.

O'Neil, A. (2005), "Nuclear nightmares", comment in the transcript of Australian Broadcasting Corporation (ABC), Radio National, Background Briefing programme, 22 May.

Organization of African Unity, Secretariat, CM/Res.1147-1176, 1988.

Roelf, W. (2005), 'SA has 'no real plan' for nuclear waste', Mail and Guardian *on line,* 25 May.

Rumley, D. (1999), *The Geopolitics of Australia's Regional Relations,* Dordrecht: Kluwer.

Rumley, D. (2005), 'The geopolitics of global and Indian Ocean energy security', in Rumley, D. and Chaturvedi, S., eds., *Energy Security and the Indian Ocean Region,* Delhi: South Asian Publishers, pp. 34-53.

Shanahan, D. (2006), 'We could be energy superpower: Howard', The Australian newspaper, 18 July, page 1.

Smith, L. (2006), 'Smuggles nuclear waste cases double', *The Weekend Australian newspaper,* 7-8 October, page 12.

The West Australian newspaper (2006), "Australia has no waste obligation', Editorial, 16 May, p. 16.

UNEP (2000), *Global Environment Outlook,* Chapter 3: Policy Responses - Africa.

United States Energy Information Administration (2004), International Energy Outlook 2004.

Uranium Information Centre (UIC) web site: http://www.uic.com.au/

World Nuclear Association web site: http://www.world-nuclear.org/

Chapter 8

Securing Shipments of Uranium and Nuclear Waste in the Indian Ocean

Vivian Louis Forbes

Introduction

This chapter has four principal aims. First, to describe the structure of SLOCs by which shipments of uranium and nuclear waste flow through the Indian Ocean. Second, to critically review the current body of international law in relation to this. Third, to discuss in detail the INF code for the safe shipment of hazardous materials adopted in 1993. Finally, the actual practice of selected Indian Ocean states in relation to this body of law and to other regional agreements will be analysed.

In general, it seems that the nuclear transportation industry has had a comparatively unblemished track record. There has never been a dangerous radioactive spill or loss recorded since the 1960s during the transportation of nuclear materials. Uranium shipments of over 50,000 tonnes have been made since the early 1970s, and by mid-2005, there had never been any reported accident in which a container with highly radioactive material had been breached, or had leaked. This is fortunate and thus speaks volumes for the professional manner by which the cargo is transported.

Freedom of navigation does exist in the Indian Ocean Region, although certain restrictions are imposed by at least eight coastal and island states. The limitations appear to be aimed at the carriage and transportation of nuclear fuel and other hazardous materials and to prohibit entry, without prior notification, of nuclear-powered ships.

Nuclear Waste Shipments in the Indian Ocean

The illustration in Figure 1, sourced from the Nuclear Control Institute, depicts four possible routes that ships have taken in order to transport nuclear material and associated hazardous cargoes from European ports to clients in Japan. The dashed blue route, potentially transiting the Indian Ocean basin, is not an option, as indicated by the statements issued by Governments of some littoral states and especially those voiced by Indonesia and Malaysia among others.

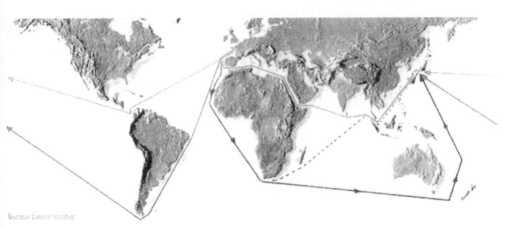

Figure One: Actual and potential sea lanes for the transportation of nuclear material.
Source: http://www.nci.org/nci-wm-sea.htm (accessed on 24 May 2005)

Green route: Sea shipment route via Panama Canal was used for 1998 waste shipment. It was also used in 1984 for first plutonium shipment.
Distance: 12,156 nautical miles (19,563 km).

Blue route: Sea shipment route via Cape of Good Hope, used for the 1997 nuclear   waste shipment from France to Japan. It was also used in 1992 for plutonium shipment (ship avoided Strait of Malacca in response to protests from Indonesia and Malaysia [route shown with dotted line] and went between Australia and New Zealand instead [solid line]).
Distance: 14,248 nautical miles (22,929 km).

Red route: Sea shipment route used in 1995 to ship waste around Cape Horn.
Distance: 16,661 nautical miles (26,813 km).

Purple route: Purple route: Sea shipment route via Suez Canal.
Distance: 10,899 nautical miles (17,540 km). Route not used to date.

Once spent fuel is removed from the nuclear reactor, it can be stored temporarily at the power plant site, transported to temporary storage off-site, or shipped to reprocessing plants. Transportation by rail or truck to interim storage facilities is normally domestic, while shipments to reprocessing sites tend to be international plants. A number of countries including Belgium, France, Germany, India, Japan, the Netherlands, Russia, Switzerland and the United Kingdom reprocess a portion of their spent fuel (Van Dyke, 1993). The main commercial re-processing/recycling facilities are based in France and the United Kingdom. Countries which send their spent fuel to France or the United Kingdom for reprocessing retain ownership of all the products, including any waste material, which must be returned to them. After shipment to the country of origin, the wastes are stored for eventual disposal. Plutonium returned as mix oxide (MOX) fuel is loaded into reactors for electricity production.

According to the Uranium Information Centre (UIC), Melbourne, about 20 million packages of all dimensions containing radioactive materials are routinely transported worldwide annually on public roads, railways and ships. The materials are placed in robust and secure containers. Transportation by sea is generally in purpose-built ships that are constructed to Classification (See INF, below).

The daily attention to shipboard safety is a reminder that nuclear materials and associated hazardous ships' cargoes are not ordinary (OECD, 1981). There are few companies in the world that specialize in this politically charged, bureaucratic task of transporting radioactive waste (Pedrozo, 1997). For those employed in the industry they would have undertaken many years of formal and informal on-the-job training. Generally, the uranium is well below weapons grade, although the cargo could still be a potential target. For this reason alone, there are volumes of international and national regulations (Kwiatkowska and Soons, 1993) designed to keep the cargo safe and out of the hands of unauthorized agents.

The final shipment of spent fuel from Australia's old HIFAR research reactor at Lucas Heights, Sydney was transported to COGEMA's La Hague plant in France for reprocessing on 22 November 2004. The cargo arrived in France on 6 January 2005 (ANSTO, 2005). It comprised 276 fuel elements, representing about seven years of operation and three million nuclear medicine treatments. The resulting wastes will be returned to Australia by 2015. This was the seventh shipment to Europe or the USA since 1963, and it leaves 469 US-origin fuel elements which will be sent to the USA in two shipments after the reactor is shut down. Commissioning of the replacement research reactor is due to begin late in 2005, and the intention is that its spent fuel will either be returned to the USA or reprocessed in France. The US Department of Energy has just extended its take-back program for foreign research reactor fuel to 2019, making specific reference to Australia's new reactor. (ANSTO, 22 Nov. 2004; 7 Jan. 2005)

In December 2002, Spanish naval forces, acting on intelligence from the United States, stopped a North Korean ship transporting missiles and warheads to Yemen. Whilst the US Government desires to set the pace on the security of the sea lanes of communication, it is generally the policy of the United States Navy to neither confirm nor deny that their ships are carrying nuclear materials. Thus, sea lanes are secured in some instances, whilst in others, there is cause for concern.

Two primary questions come to the fore in addressing the topic of security of the sea lanes of communication (SLOCs) within the Indian Ocean basin. First, do the littoral states have the right to impose restrictions on the types of ships and the nature of cargo and/or purpose of activity that all types of marine craft may be involved in? Second, does freedom of navigation exist, and if so, to what extent, especially in the Indian Ocean Basin? Shipments of nuclear cargo and other hazardous material have often attracted wide attention and controversy, particularly among the littoral states along the routes traversed by the ships carrying such commodities. Other concerns and questions raised include: those related to the safety and physical security of the shipments; questions on the impact on the marine environment in the unlikely event of an accident; issues associated with the liability of the shipping states; and other unforeseen circumstances.

Within the semi-enclosed seas of Southeast Asia, which in essence comprise in the eastern portion of the Indian Ocean Basin the regional grouping referred to as the Association of Southeast Asian Nations (ASEAN), an agreement was signed to declare their maritime area a Zone of Peace, Freedom and Neutrality (ZOPFAN) and in a declaration known as the Treaty of Bangkok made provision with reference to the perceived rights of navigation of foreign-flagged ships. The South East Asian Nuclear Weapon-Free (*NWF*) Zone Treaty (*Treaty of Bangkok*) opened for signature in 1995 and came into force in 1996. No nuclear weapon state has signed the additional protocol, which calls for respect for the Treaty and security assurances.

Article 7 of the Bangkok Treaty (ZOPFAN) relating to movement of foreign ships and aircraft specifies that:

> Each State Party, on being notified, may decide for itself whether to allow visits by foreign ships and aircraft to its ports and airfields, transit of its airspace by foreign aircraft, and navigation by foreign ships through its territorial sea or archipelagic waters and over flight of foreign aircraft above those waters in a manner not governed by the rights of innocent passage, archipelagic sea lanes passage or transit passage.

International Law and Sea Lanes of Communication

Freedom of Navigation

In a seminal dissertation in 1608, Dutch legal scholar, Hugo Grotius, argued that navigation should be free to all (Magoffin, 1916), and that innocent passage on the seas should not be curtailed (Mare Liberum). Grotius was of the opinion that the Dutch possessed rights to trade in the Indian Ocean Region and the East Indies archipelago, which the Portuguese, in that era, perceived to be their exclusive domain (Forbes, V., 1995, p. 42). There was a counter-argument which sought a closed sea approach (Mare Clausum). The principle of the freedom of the seas was enshrined, nearly four centuries later, in the 1982 United Nations Convention of the Law of the Sea, (1982 Convention) and stated in Article 87 of the said Convention. Underlining the profound importance of the seas to the international community (Van Dyke, 2002), the 1982 Convention established a comprehensive international framework to facilitate and promote the peaceful uses of sea lanes of communication (SLOCs).

The concept of freedom of navigation continues to be of immense concern and importance in 2005, and indeed, will in the future. The oceans, sea and straits remain a geographical setting for the livelihood of millions of people, for the prosperity of countries around the world, as well as a key source of marine biotic and mineral resources. It is therefore extremely important that the seas remain free and safe to navigate upon and within, and this is particularly true for the Indian Ocean basin.

During the 1600s, only a handful of maritime states possessed ships capable of voyaging to distance lands for the purposes of exploration, empire-building and the development and expansion of trade. In 2005, almost every state - coastal, island and land-locked states - is involved in maritime trade either directly or indirectly. Even with advances in other forms of communications and transportation, maritime activity continues to increase (Forbes, A., 2004). Indeed, the United Nations Conference on Trade and Development predicted that world maritime traffic would grow by up to five per cent annually over the next ten years (UNCTAD Report, 2003).

In the Asia-Pacific region, this maritime trade is important not just to its littoral states, but to world commerce. The Straits of Malacca and Singapore and the numerous seas and straits within the Indonesian archipelago are among the busiest waterways in the world. There are also 'constricted sea lanes of communication' in the Red Sea, Persian Gulf, Mozambique Channel and the Torres Strait. All of these geographical features are located strategically to guard the Indian Ocean basin. Within these semi-enclosed seas, zones of maritime jurisdiction have been delimited or are in the process of being defined (Forbes, V., 1995 and 2001).

The underlying issues in the controversy of freedom of the sea, and especially that of nuclear-powered ships, have often centred on the fundamental debate between the rights of littoral states to apply the principle of mandatory prior-informed consent and the precautionary principle law (Van Dyke, 1996). These relate to the rights of maritime states to freedom of navigation and innocent passage (Article 21), in accordance with what has been argued as a widely-accepted rule in international customary passage through the territorial seas and archipelagic waters, as well as through the international straits and the Exclusive Economic Zones of transit states, as provided for under the 1982 Convention. Indeed, in the context of nuclear-powered ships, the provision in Article 23 is clear:

> Foreign nuclear-powered ships and ships carrying nuclear or other inherently dangerous or noxious substances shall, when exercising the right of innocent passage through the territorial sea, carry documents and observe special precautionary measures established for such ships by international agreements.

Article 25, paragraph 3, of the 1982 Convention stipulates that a coastal state may, without discrimination in form or in fact among foreign ships, suspend temporarily, in specified areas of its territorial sea the innocent passage of foreign ships, if such suspension is essential for the protection of its security, including weapons exercises. Such suspension takes effect, according to the same article, only after having been duly published.

The maritime states contend that the provisions in the 1982 Convention guarantee freedom of navigation through: the high seas (Article 87); the exclusive economic zones (Article 58:1); straits used for international navigation (Articles 34 to 44); and archipelagic sea lanes (Article 53). They further argue that the 1982 Convention also permits the exercise of 'innocent passage' through the territorial seas (Articles 17 to 19), and through archipelagic waters (Article 52).

Littoral and transit states, on the other hand, maintain that the provisions in the 1982 Convention also impose an obligation on all states to protect and preserve the marine environment (Article 192), and to take measures to prevent, reduce and control pollution of the marine environment (Article 194). These states also contend that the 1982 Convention recognises the specific powers of coastal and island states to regulate ships carrying nuclear or other inherently dangerous or noxious substances or materials (Articles 22 and 23), as well as the duty of all states to notify other affected states immediately in cases where 'the marine environment is in imminent danger of being damaged or has been damaged by pollution', as provided for in Article 198.

Another argument is whether the passage of vessels carrying irradiated nuclear fuel (INF) or highly radioactive waste can be considered as innocent as defined under Article 23 of the 1982 Convention, (Marin, 2001) if such a passage did not comply with special precautionary measures established for ships by international agreements. A resemblance of such an agreement is the Code for the Safe Carriage of Irradiated Nuclear Fuel, Plutonium and High-Level Radioactive Wastes in Flasks on Board Ships (INF Code), which was adopted by the International Maritime Organization (IMO) in 1993.

INF Code

A recommendatory 'Code for the Safe Carriage of Irradiated Nuclear Fuel, Plutonium and High-Level Radioactive Wastes in Flasks on board Ships' (INF Code) was adopted by the eighteenth session of the IMO Assembly on 4 November 1993 (Resolution A. 748 (18)). The 20th session of the Assembly adopted amendments to this INF Code to include specific requirements for shipboard emergency plans and notification in the event of an incident (Resolution A. 853 (20) which was adopted on 27 November 1997).

The International Code for the Safe Carriage of Packaged Irradiated Nuclear Fuel, Plutonium and High-Level Radioactive Wastes on Board Ships (INF Code) was adopted by resolution MSC.88 (71) on 27 May 1999 and became mandatory on 1 January 2001 by amendments to Chapter VII of SOLAS (Carriage of dangerous goods). It set out how the material covered by the Code should be carried, including specifications for ships. The material covered by the code includes: irradiated nuclear fuel - material containing uranium, thorium and/or plutonium isotopes which has been used to maintain a self-sustaining nuclear chain reaction; plutonium - the resultant mixture of isotopes of that material extracted from irradiated nuclear fuel from reprocessing; and high-level radioactive wastes - liquid wastes resulting from the operation of the first stage extraction system or the concentrated wastes from subsequent extraction stages, in a facility for reprocessing irradiated fuel, or solids into which such liquid wastes have been converted. The INF code applies to all ships regardless of the date of construction and size, including cargo ships of less than 500 gross tonnage, engaged in the carriage of INF cargo. However, it does not apply to warships, naval auxiliary or other ships used only on government non-commercial service, although administrations are expected to ensure such ships are in compliance with the Code. Specific regulations in the Code cover a number of issues, including: damage stability, fire protection, temperature control of cargo spaces, structural consideration, cargo securing arrangements, electrical supplies, radiological protection equipment and management, training and shipboard emergency plans. Ships carrying INF cargo are assigned to one of three classes, depending on the total radioactivity of INF cargo which is carried on board, and regulations vary slightly according to the Class:

- Class INF 1 ship - Ships which are certified to carry INF cargo with an aggregate activity of less than 4,000 TBq (TerrBecquerel is a unit of measurement of radioactivity);

- Class INF 2 ship - Ships which are certified to carry irradiated nuclear fuel or high-level radioactive wastes with an aggregate activity of less than 2×10^6 TBq and ships which are certified to carry plutonium with an aggregate activity less than 2×10^5 TBq; and,

- Class INF 3 ship - Ships which are certified to carry irradiated nuclear fuel or high-level radioactive wastes and ships which are certified to carry plutonium with no restriction of the maximum aggregate activity of the materials.

The INF Code established certain physical standards for the carriage of nuclear and radioactive material, but the IMO Assembly Resolution that adopted the Code had also explicitly acknowledged the omission of considerations of emergency response planning, notification of coastal states on the shipment of nuclear materials, positive tracking of vessels transporting INF2 and INF3 quantities, and of the fitting of transport containers with devices to assist their location and recovery, should they be lost at sea, in adopting the Code. The same IMO Assembly Resolution also noted the need for an adequate liability and compensation regime for damage in connection with the carriage by sea of radioactive matter, and, in that regard, requested the Organization, in consultation with the International Atomic Energy Agency (IAEA), to review the INF Code, with a view to augmenting or improving on it.

The INF Code of SOLAS is but one of many IMO documents that sets the standards for operations in maritime trade. SOLAS 1974, which entered into force 25 May 1980, has 155 contracting states - representing some 98.52 per cent of world tonnage (IMO, April 2005). The SOLAS Protocols of 1978 and 1988 have

107 and 77 contracting states respectively (94.99 and 66.84 per cent of world tonnage). Of those states in the Indian Ocean Region, 32 are contracting members to all or part of SOLAS; Somalia is not listed.

With reference to the Convention Relating to Civil liability in the Field of Maritime Carriage of Nuclear Material, 1971 (see Appendix, below) and the carriage of hazardous and Noxious Substances by Sea (NHS), 1996 there are no contracting states from the IOR.

Select IOR States' Practice

Let us now examine the declarations made by a selection of eight IOR states - Bangladesh, Egypt, India, Iran, Malaysia, Oman, Pakistan and Yemen - with particular reference to the provisions of the 1982 Convention relating to freedom of navigation and the carriage of nuclear materials when these states became party either on signature and/or at the time of lodging their instruments of ratification.

Since the mid-1970s, the littoral states of the Indian Ocean basin have claimed extended maritime jurisdiction and additional zones thereby pushing the limit of the 'high sea' to a distance of 200 nautical miles from the coast of coastal and island states. Collectively, nearly 32 per cent (22 million square kilometers) of the surface area of the Indian Ocean falls within national jurisdiction (the EEZs of the states) (Forbes, V., 1995, p. 103). The remaining 68 per cent is generally considered 'free for navigation' to the international community, although 'innocent passage' and 'freedom of navigation' are perceived to be guaranteed in the various national jurisdictional zones within the letter and spirit of international law.

The declaration by Bangladesh upon ratification of the 1982 Convention on 27 July 2001 stated:

> 3. The exercise of the right of innocent passage of warships through the territorial sea of other states should also be perceived to be a peaceful one. Effective and speedy means of communication are easily available and make the prior notification of the exercise of the right of innocent passage of warships reasonable and not incompatible with the Convention. Such notification is already required by some states. Bangladesh reserves the right to legislate on this point.

> 4. Bangladesh is of the view that such a notification requirement is needed in respect of nuclear-powered ships or ships carrying nuclear or other inherently dangerous or noxious substances. Furthermore, no such ships

shall be allowed within Bangladesh waters without the necessary authorization.

Egypt's declaration concerning the passage of nuclear-powered and similar ships through the territorial sea of Egypt was made clear on 26 August 1983. It stated:

> Pursuant to the provisions of the Convention relating to the right of the coastal state to regulate the passage of ships through its territorial sea and whereas the passage of foreign nuclear-powered ships and ships carrying nuclear or other inherently dangerous and noxious substances poses a number of hazards,
> Whereas Article 23 of [the 1982] Convention stipulates that the ships in question shall, when exercising the right of innocent passage through the territorial sea, carry documents and observe special precautionary measures established for such ships by international agreements.

The Government of the Arab Republic of Egypt declared that it will require the aforementioned ships to obtain authorization before entering the territorial sea of Egypt, until such international agreements are concluded and Egypt becomes a party to them. In its declaration concerning the passage of warships through the territorial sea of Egypt, it stressed that warships shall be ensured innocent passage through the territorial sea of Egypt, subject to prior notification. A specific declaration concerning passage through the Strait of Tiran and the Gulf of Aqaba noted that the provisions of the 1979 Peace Treaty between Egypt and Israel concerning passage through the Strait of Tiran and the Gulf of Aqaba would come within the framework of the general regime of waters forming straits referred to in Part III of the Convention, wherein it is stipulated that the general regime shall not affect the legal status of waters forming straits and shall include certain obligations with regard to security and the maintenance of order in the state bordering the strait. Egypt also stated that it would exercise those rights attributed to it by the provisions of Parts V and VI of the 1982 Convention in the EEZ situated beyond and adjacent to its territorial sea in the Mediterranean Sea and in the Red Sea.

India's declaration was made upon ratification of the 1982 Convention on 29 June 1995. It stated in part:

> (b) The Government of the Republic of India understands that the provisions of the Convention do not authorize other states to carry out in the exclusive economic zone and on the continental shelf military exercises or manoeuvres, in particular those involving the use of weapons or explosives without the consent of the coastal state.

Iran's declaration, which was made on the opening day, 10 December 1982, of the adoption of the 1982 Convention was emphatic, in noting that:

> 2) In the light of customary international law, the provisions of article 21, read in association with article 19 (on the Meaning of Innocent Passage) and article 25 (on the Rights of Protection of the Coastal States), recognize (though implicitly) the rights of the Coastal States to take measures to safeguard their security interests including the adoption of laws and regulations regarding, inter alia, the requirements of prior authorization for warships willing to exercise the right of innocent passage through the territorial sea.

A statement by Malaysia on the occasion of depositing its instrument of ratification of the 1982 Convention on 14 October 1996, read in part:

> 4. In view of the inherent danger entailed in the passage of nuclear-powered vessels or vessels carrying nuclear material or other material of a similar nature and in view of the provision of article 22, paragraph 2, of the Convention on the Law of the Sea concerning the right of the coastal State to confine the passage of such vessels to sea lanes designated by the State within its territorial sea, as well as that of article 23 of the Convention, which requires such vessels to carry documents and observe special precautionary measures as specified by international agreements, the Malaysian Government, with all of the above in mind, requires the aforesaid vessels to obtain prior authorization of passage before entering the territorial sea of Malaysia until such time as the international agreements referred to in article 23 are concluded and Malaysia becomes a party thereto. Under all circumstances, the flag state of such vessels shall assume all responsibility for any loss or damage resulting from the passage of such vessels within the territorial sea of Malaysia.

Oman's Declaration No. 3, of 7 August 1989, relates to the passage of foreign nuclear-powered ships or those vessels carrying nuclear or other substances that are inherently dangerous or harmful to health or the environment. The right of innocent passage, subject to prior permission, is guaranteed to the types of vessel, whether or not warships, to which the descriptions apply. This right is also guaranteed to submarines to which the descriptions apply, on condition that they navigate on the surface through Omani territorial waters and fly the flag of their home state.

The Government of the Islamic Republic of Pakistan in its process of ratifying the 1982 Convention declared, on 26 February 1999, in part:

(ii) The Law of the Sea Convention, while dealing with transit through the territory of the transit State, fully safeguards the sovereignty of the transit state. Consequently, in accordance with Article 125, the rights and facilities of transit to the land-locked state ensure that it shall not in any way infringe upon the sovereignty and the legitimate interest of the transit state. The precise content of the freedom of transit consequently, in each case, has to be agreed upon by the transit state and the land-locked state concerned. In the absence of such an agreement concerning the terms and modalities for exercising the right of transit, through the territory of the Islamic Republic of Pakistan shall be regulated only by national laws of Pakistan;
(iii) It is the understanding of the Government of the Islamic Republic of Pakistan that the provisions of the Convention on the Law of the Sea do not in any way authorize the carrying out in the exclusive economic zone and in the continental shelf of any coastal state military exercises or manoeuvres by other states, in particular where the use of weapons or explosives is involved, without the consent of the coastal State concerned.

The eighth State, Yemen, made its declaration on 21 July 1987, and noted that:

1. The People's Democratic Republic of Yemen will give precedence to its national laws in force which require prior permission for the entry or transit of foreign warships or of submarines or ships operated by nuclear power or carrying radioactive materials.

Nuclear Weapon Free Zones
Nuclear Weapon Free Zones (NWFZs), in concept, prohibit the stationing, testing, use, and development of nuclear weapons inside a particular geographical region, whether that is a single state, a region, or an area governed solely by international agreements. They have been identified in many international forums, including the Non-Proliferation Treaty (NPT) and the UN General Assembly, as being positive steps towards nuclear disarmament (Masperi, 1997).

In the Middle East, the establishment of a zone free of Israel's nuclear weapons, and all other weapons of mass destruction would be a key component of regional security. In Central Asia, the emergence of a zone treaty, which seemed imminent, now faces some political obstacles. These need to be overcome. In volatile South

Asia, which witnessed a nuclear breakout with the Indian and Pakistani tests of 1998, an NWFZ could prevent the deployment of nuclear weapons. In Northeast Asia, with Japan and North and South Koreas, an NWFZ would offer the best guarantee of security without nuclear weapons while ensuring that no country crosses the nuclear threshold.

Single State Zones (SSZ) are created by national legislation, declaration or constitutional mandate. Some countries, such as those of New Zealand and the Philippines, have used domestic means to go beyond the obligations of the regional treaty to which they are a party. New Zealand's Nuclear Free Zone domestic legislation prohibits any foreign ship that is nuclear-powered or carrying nu clear weapons from entering its internal waters or any foreign aircraft landing in its territory. This goes beyond New Zealand's obligations under the South Pacific Nuclear Free Zone Treaty, which permits port visits of nuclear ships. The Philippines, a member of the South East Asian Nuclear Weapon Free Zone, has declared its territory free of nuclear weapons through a change in its constitution.

An Indian Ocean Case Study

The Indian Supreme Court monitoring committee on hazardous waste recommended that the Danish ship, formerly King Frederik IX, should be 'driven out' of Indian waters. It noted, furthermore, that an enquiry should be set up to identify those responsible for the vessel's illegal entry. The Gujarat Maritime Board, Gujarat Pollution Board and Central Pollution Board, had inspected the ship and gave it permission to be beached, but not broken. Permitting the ship to stay in Indian maritime jurisdictional space could have far-reaching and adverse implications to that country's environmental care and concerns and international image. The committee was expected to pass a formal order on 2 June 2005 for the ship to return to Denmark for decontamination and only thereafter would it be permitted to return to India in accordance with the Basel Convention as well as the regulations of the Indian Government.(*Fairplay*, Lloyds, 31 May 2005).

Conclusion

The nuclear transportation industry has had a comparatively unblemished track record. There has never been a dangerous radioactive spill or loss recorded since the 1960s during the transportation of nuclear materials. Uranium shipments of over 50,000 tonnes have been made since the early 1970s, and, by mid-2005, there has never been any accident in which a container with highly radioactive material had been breached, or had leaked. This is fortunate and thus speaks volumes for the professional manner by which the cargo is transported.

Freedom of navigation does exist in the Indian Ocean Region, although, as we have seen above, certain restrictions are imposed by at least eight coastal and island states. These limitations appear to be aimed at the carriage and transportation of nuclear fuel and other hazardous materials and to prohibit entry, without prior notification, of nuclear-powered ships.

References

ANSTO (2004), 'Final HIFAR spent Fuel Shipment for France Leaves', Australian Nuclear Science and Technology Organisation, *Media Release*, 22 November.

ANSTO (2005), 'Spent Fuel Arrives Safely in France', *Media Release*, 6 January 2005.

Forbes, Andrew (2004), 'International Shipping: Trends and Vulnerabilities', Paper presented at the *Conference on Maritime Security in Asia,* Hawaii, 18-20 January.

Forbes, V. L. (1995), *The Maritime Boundaries of the Indian Ocean Region*, Singapore: Singapore University Press.

Forbes, V. L. (2001), *Conflict and Cooperation in Managing Maritime Space in Semi-enclosed Seas*, Singapore: Singapore University Press.

International Atomic Energy Agency Waste Management Section, Division of Nuclear Fuel Cycle, (1993), 'Technical Paper in Support of the Code of Practice on Transboundary Shipments of Nuclear Wastes' (April 1989), reprinted in Barbara Kwiatkowska & Alfred Soons (eds.), *Transboundary Movements and Disposal of Hazardous Wastes in International Law* Dordrecht: M. Nijhoff Publishers, pp 355-392.

Kwiatkowska, Barbara & Soons, Alfred (eds.), (1993), *Transboundary Movements and Disposal of Hazardous Wastes in International Law*, Dordrecht: M. Nijhoff Publishers.

Maggoffin, R van Deman (1916), 'A translation of H. Grotius', (1606), *The Freedom of the Seas*, New York: Oxford University Press.

Marin, Lawrence (2001), 'Oceanic Transportation of Radioactive Materials: The Conflict Between the Law of the Seas' Right of Innocent Passage and Duty to the Marine Environment', *Florida Journal of International Law*, 13: 361.

Masperi, Luis (1997), 'Present and Future Nuclear-Weapon-Free Zones', *Bulletin* 10 - Towards a Nuclear-Weapon-Free-World.

OECD Nuclear Energy Agency, (1981), *Nuclear Legislation: Analytical Study: Regulations Governing the Transport of Radioactive Materials*.

Pedrozo, R. A. F., (1997), 'Transport of Nuclear Cargoes by Sea', *Journal of Maritime Law and Commerce*, 28: 207.

UN Conference on Trade and Development (2003), *Trade and Development Report*, New York: United Nations.

Uranium Information Centre (2003), 'Transport of Radioactive Materials', *Nuclear Issues Briefing Paper* No. 51.

Van Dyke Jon M., (1993), 'Sea Shipment of Japanese Plutonium under International Law', *Ocean Development and International Law,* 24: 399.

Van Dyke, Jon M. (1996), 'Applying the Precautionary Principle to Ocean Shipments of Radioactive Materials', *Ocean Development and International Law*, 27: 379.

Van Dyke, Jon M. (2002), 'The Legal Regime Governing Sea Transport of Ultra-hazardous Radioactive Materials', *Ocean Development and International Law*, 33: 77-108.

Appendix

Convention relating to Civil Liability in the Field of Maritime Carriage of Nuclear Material (NUCLEAR), 1971

Adopted: 17 December 1971   Entry into force: 15 July 1975

In 1971, the IMO, in association with the International Atomic Energy Agency (IAEA) and the European Nuclear Energy Agency of the Organization for Economic Co operation and Development (OECD), convened a Conference which adopted a Convention to regulate liability in respect of damage arising from the maritime carriage of nuclear substances.

The purpose of this Convention is to resolve difficulties and conflicts which arise from the simultaneous application to nuclear damage of certain maritime conventions dealing with ship-owners' liability, as well as other conventions which place liability arising from nuclear incidents on the operators of the nuclear installations from which or to which the material in question was being transported.

The 1971 Convention provides that a person otherwise liable for damage caused in a nuclear incident shall be exonerated for liability if the operator of the nuclear installation is also liable for such damage by virtue of the Paris Convention of 29 July 1960 on Third Party Liability in the Field of Nuclear Energy; or the Vienna Convention of 21 May 1963 on Civil Liability for Nuclear Damage; or national law which is similar in the scope of protection given to the persons who suffer damage.

Chapter 9

Indian Ocean Gateway: Global LNG Trade Movement

Noor Apandi Osnin and Siti Norniza Zainul Idris

Introduction

Natural Gas is the world's third largest primary energy source after coal and oil. More than four-fifths of the world production is consumed locally, with the rest traded internationally either by pipeline or on liquid natural gas (LNG) ships. Natural gas, with its higher thermal efficiency and low emission of environmental pollutants has become the future fuel of choice worldwide. The industry started in 1964, with the first commercial trade taking place between Algeria and United Kingdom. This has placed the United Kingdom as the world's first LNG importer and Algeria as the first LNG exporter. Following the success of the first LNG trade, additional liquefaction plants and terminals were constructed in both basins, the Atlantic, the Pacific, as well as the Middle East. As a result, a growing share of gas traded between these regions either by pipeline or LNG has increased steadily over the years. The industry expanded and this is demonstrated by the increase in gas production and consumption by each region. This chapter will highlight the expansion of the natural gas industry throughout the world as well as the emergence of the Indian Ocean as one of the gateways for LNG transshipment. In addition, the safety aspects of LNG are also examined.

An Introduction to Liquid Natural Gas (LNG)

Natural gas, as well as oil, is found under the surface and seas at different depths and levels. It was previously considered as a nuisance by-product of crude oil production until the 20th century (Noor Apandi Osnin, 2004). Natural gas is a cryogenic[1], colourless, odourless and non-toxic liquid compared to other gases. It is an abundant fossil composed primarily of methane, hydrocarbons (ethane, propane and butane) and other inert gases such as nitrogen and carbon dioxide. In order for natural gas to be available for trade and for local use, the gas needs to be condensed into a liquid at approximately -161° C and at atmosphere pressure before it can be piped into insulated storage tanks for distribution by sea-going ships and pipelines. The liquefaction process removes non-methane components such as water, carbon dioxide, sulfur and hydrocarbons from the natural gas and reduces the gas volume to make it more economical for transportation and storage. There are four main stages in the LNG process, as follows:

- Exploration of the natural gas, which most of the time is discovered during the search for oil;
- Liquefaction process - this is where the natural gas is converted into liquid form with low temperature approximately -161¡C;
- Shipping, transfer of LNG in its vessels; and

- Regasification and storage, from the liquefied form to the gaseous form and then the LNG is stored in specially made storage tanks.

A question has been raised over why LNG is considered as an alternative energy source. Compared to other energy sources - for example oil, petroleum, coal and diesel - natural gas is considered to be more efficient and environmental friendly. It offers an energy density compared to petrol and diesel fuels and produces less pollution. The minimal impact from its production, shipment and regasification on the environment and human health are factors that have increased the demand for natural gas. Furthermore, natural gas produces minimum toxic emissions of carbon dioxide and other greenhouse gases such as sulphur dioxide and nitrous oxide. In countries with a serious pollution problem, LNG is considered as an alternative source of energy as it is a reliable and safe energy source compared to oil and coal. There has been a great demand for LNG as more players are taking part in LNG investment. Cost reductions in operating and transportation have significantly declined due to the emergence of new technologies in the industry. Furthermore, due to the price increase in oil supplies, natural gas has become an economic alternative. With the expansion in the number of LNG carriers and tonnage, LNG is expected to grow more in the next seven years than it has done in the previous 40 years (Spero News, 2005). The industry is forecasted to have a substantial growth from 150% to over 210% to 2020.

Supply and Demand in the LNG Market

An Overview of the LNG Industry

- The natural gas industry is the fastest growing energy industry. The industry is projected to increase by 2.7 - 3.0 % per year.
- 12 countries operate liquefaction facilities, including Abu Dhabi, Algeria, Australia, Brunei, Indonesia, Libya, Malaysia, Nigeria, Oman, Qatar, Trinidad and Tobago and United States.
- 10 consuming countries operate 32 LNG terminals mainly in Japan (23 terminals), three in Spain, three in United States, two in France and one each in Belgium, Greece, Italy, Taiwan, China and Turkey.
- Seven additional countries - Angola, Bolivia, Equatorial Guinea, Iran, Peru, Venezuela and Yemen - have the potential to be natural gas suppliers.
- China, India and United Kingdom are expected to become the primary LNG importers, replacing Japan as the major importer.
- The largest growth in production is projected for the Middle East, 8.3 Tcf [2] in 2001 up to 18.8 Tcf in 2025. Russia's (EE/FSU [3]) production exceeded nearly 15.0 Tcf for the year 2025. The total production by these two regions has put the Middle East and Russia as the major sources of supply.
- Based on the different supply and demand profiles of the Atlantic and Pacific Basins, the Middle East will be the pivot point that balances the two markets [4].

LNG Supply and Demand

The natural gas industry is projected to increase by an average of 2.7- 3.0% annually from the year 2001 up to 2025. This projection represents a robust growth in the natural gas industry in becoming the primary key for the world energy industry. The high value market in the Atlantic Basin, Middle East and the Pacific Basin encourages an increase in volume capacity to satisfy short and long-term demand by the consuming countries.

The increase in the world natural gas trade is based on trade consumption by each supplier. The Atlantic Basin and Middle East are well positioned to support the existing LNG market worldwide. A number of existing projects such as Rasgas and Qatargas II in Qatar, Bontang and Arun Project in Indonesia, Bintulu LNG complex in Malaysia and North West Shelf Project in Australia are scheduled for expansion. With new players coming in, the industry is expected to become bigger and provide a more competitive market for supplier countries.

An Australian Bureau Economics of Agriculture and Resource Economics (ABARE) study projected that Qatar would sell most of its gas production to Europe and to the US east coast gas market[5]. There is a potential for the US and the European countries to rely heavily on the Middle East supplies as their primary source. This is due to low cost gas reserves in the Middle East and the large production in LNG volume compared to the Atlantic and Pacific Basins.

Table 1
World Natural Gas Production by Region, 2001-2025 (in Tcf)

Region/Country	Projections					Percent Change, 2001-2025
	2001	2010	2015	2020	2025	
• North	27.6	29.6	30.6	32.8	33.6	0.8
America	19.7	20.5	21.6	23.8	24.0	0.8
• United States	6.6	7.6	7.5	7.1	7.5	0.5
• Canada	1.3	1.5	1.6	1.9	2.1	2.0
• Mexico	10.2	9.0	9.0	8.9	9.8	-0.2
• Western Europe						
Europe	0.1	0.1	0.1	0.1	0.1	-1.0
• Japan	1.4	2.3	2.9	3.1	3.4	3.7
• Australia/New Zealand						
Total	**39.3**	**40.9**	**42.6**	**44.9**	**46.8**	**0.7**
• Russia	25.7	30.2	34.9	39.6	44.5	2.3
• Eastern Europe	0.9	0.9	0.8	0.8	0.8	-0.5
Total	**26.6**	**31.0**	**35.7**	**40.4**	**45.3**	**2.2**
• China	1.1	1.6	1.9	2.3	3.1	4.5
• India	0.8	0.9	0.9	1.2	1.5	2.6
• South Korea	0.0	0.0	0.0	0.0	0.0	-
• Other	6.9	7.7	8.3	9.6	10.8	1.9
• Middle East	8.3	9.8	12.1	15.6	18.8	3.5
• Africa	4.6	8.1	9.9	11.9	14.1	4.8
• Central and South America	3.6	5.5	7.1	8.6	10.6	4.6
Total	**25.2**	**33.5**	**40.2**	**49.2**	**58.9**	**3.6**
Total World	**91.1**	**105.5**	**118.5**	**134.5**	**151.0**	**2.1**

Source : Energy Report. www.eia.doe.gov

Table 1 shows that the largest growth in production is projected for the Middle East - 8.3 Tcf[6] in 2001 to 18.8 Tcf in 2025. The percentage of annual projection for the Middle East production is 3.5%. Russia's production exceeded 44.5 Tcf for 2025, increasing by 18.8 Tcf from 2001. However, the percentage of annual production by Russia is lower compared to the Middle East with only 2.3%. This indicates that the Middle East and Russia are the major sources for LNG supplies. The indication is also based on the total reserves for each region as illustrated in Figure 2.

Figure 2
World Natural Gas Reserves by Region

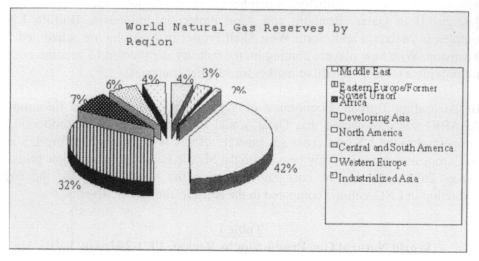

Source: Energy Outlook. www.eia.doe.gov

The increase in world natural gas demand is based on the total gas reserves of each exporter country. The newly-discovered areas hold greater prospects for an increase in demand as there is huge potential for large exploration in the near future, as many areas have not been fully explored. Based on Figure 2, the Middle East holds the largest gas reserves with 42% of the world's reserves, followed by Russia with 32%. Based on this percentage, the Middle East is expected to lead exporter countries in supplying natural gas for the next five years onwards (refer to annual projection in Table 1).

The increase in the current supply and demand of LNG has also significantly improved market competitiveness. The industry is now going through a period of massive expansion and change due to large-scale demand growth by new import players and the emergence of new import markets to replace the original players since 1964. The expansion of these two factors has brought to another level of contract feasibility between suppliers and consumers as there is a high demand for both short-term and long-term contracts.

Review of new markets and supplies
New exporters will boost the industry (Iran, Egypt, Russia, Equatorial Guinea, Yemen, Angola and Venezuela) in providing more choice and diversity in supplies. In addition, several European and US gas companies are proposing a project to pipe the LNG gas from Bolivia to either Peru or Chile where it will be liquefied and shipped to the West Coast of North America.[7]

The industry forecast that there will be another 25-35 new projects to boost the industry in 2012, which includes projects in Qatar, Algeria, Sakhalin in Russia, Melkoye Island in Norway, Egypt, Malaysia, Australia, Brunei, Indonesia, Nigeria and others.[8]   Huge investments by gas and oil companies such as BP, Chevron Texaco, ExxonMobil, Shell and Marathon Oil have increased the number of export plants in these countries, which provide more diversity of supplies and market and this has increased the resources capacity to 2012.

With this expansion, global prices are likely to change along with the rapid demand by the main importers such as the United States, the Asia-Pacific (Japan, South Korea and Taiwan) and the European markets. The emergence of new importers such as the United Kingdom, India, China, Portugal and Republic of Dominica has increased their import capacity either by constructing new terminals or by expanding the existing terminals to meet the requirements for their industrial use.

However, most of the supply and demand analysis forecasts the development of the Middle East as the primary supplier of the LNG industry. The expansion of the Middle East market will have implications for the other importers of LNG. This is triggered by the geographical location of the Middle East in the Indian Ocean, which is naturally an alternative for the Asia-Pacific and the Atlantic Basin to secure their LNG supplies.

The growing demand from the Middle East countries, mainly Qatar, is due to the new entrants of India and China in the market. Both countries are expected to experience an overall energy demand growth as both countries are going to generate huge markets over the coming decade. This will place both countries as the largest LNG importers along with Japan and the United States.

LNG Transshipment

LNG Transport
The first commercial transport of LNG occurred in 1952 from Louisiana to Chicago by a 6,000 cu. m barge called the Methane (Noor Apandi Osnin, 2004). Since then, large-scale commercial shipments began in 1964 with the first international shipment from Algeria to the United Kingdom. This attempt marked a turning point in LNG transshipment history as the number of LNG carriers has grown very rapidly ever since.

As at March 2006, there were 196 LNG vessels with a capacity of 23,889,000 cu. m in operation (Drewry Shipping Insight, 2006). There has also been an increase in the total LNG order book. In July 2005, about 115 vessels were to be delivered by 2010.

Table 2
LNG Order book by year of delivery.

Year	2005	2006	2007	2008	2009	2010
Total of Vessels	13	25	34	33	9	1
Capacity (cu.m)	1,712,500	3,519,930	5,108,900	5,313,000	1,377,000	153,000

Source: LNG Newbuilding in a nutshell. Gard News Issue, November 2005-January 2006. pg 7 & 8.

Table 2 clearly illustrates the rising demand for LNG vessels over the next four years. Gard News has reported that there are rumors that one of the LNG major suppliers is due to order 42 new vessels for delivery in the coming years.[9] There is a possibility for the number to be increased due to diversity of trade, especially demand for LNG from Europe, the United States and Asia-Pacific. With the increase in ship size and capacity in the LNG sector nowadays, it is possible now for a vessel to carry more gas compared to the traditional steam turbine design vessel. Today, most of the shipyards have likely received new orders for 250,000 cu. m vessels compared to a standard size for LNG carrier of 140,000 cu. m. It was reported that Qatargas ordered two 216,000 cu. m ships to be delivered in 2008. [10]

In addition, the price for a LNG carrier has not been increased enough to deter ship owners from ordering and buying new vessels. The Clarkson reported that the price for a 145,000 cu. m vessel in early 2005 was $185 million compared with $155 million in 2004. Moreover, the trade flow is also expanding between supplier and consumer countries.

New Route: LNG Trade Movement in the Indian Ocean

Figure 1
LNG Trade Movement from 2002-2012

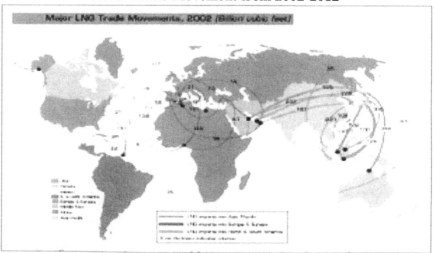

LNG Supply countries: Algeria, Libya, Trinidad and Tobago, US, Australia, Brunei, Indonesia, Malaysia, Nigeria, Oman and Qatar.
Source: The Global Liquefied Natural Gas Market: Status and Outlook. www.eia.doe.gov

Figure 1 above illustrates the movements of LNG trade from supplier countries to consuming countries in 2002. From the above, the LNG movement shows the regional trend in supply and demand of the LNG market. A large demand can be seen in the Atlantic Basin followed by the Pacific Basin and Middle East. Even though the trade movement in the Pacific is lower compared to the Atlantic Basin, the Pacific countries supplied nearly half of the global export (49%). In addition, Middle East exports 23% and the Atlantic Basin 29% of total global exports.

Recent trends in LNG trade movement have shown a change in trade flows from traditional intra-regional trade to inter-regional trade. This is due to the emergence of the Indian Ocean as a route for LNG transshipment in these three regions. The emergence of the Indian Ocean as one of the trading routes is based on the latest developments in the LNG sector in the Middle East - mainly Qatar, Oman and Abu Dhabi in the United Arab Emirates (UAE). With the expansion of the Middle East market, the Indian Ocean is expected to become the main route for LNG transshipment between the Middle East and the Asian market. The Middle East will act like a watershed for both basins. Changes in trading routes are based on the demand by consuming countries in the short and long-term (demand involved with new liquefaction project which have not been completed). LNG trade movement in the Indian Ocean can be seen below.

Figure 2
LNG Trade Movement in Indian Ocean (based on Trillion cubic metres)

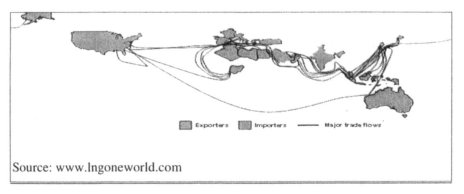

Source: www.lngoneworld.com

To simplify the trade movement as illustrated in Figure 2, the trade movement can be seen as follows:

- Middle East - Egypt to France, USA, Spain and Italy;
  - Iran to China and India;
  - Libya to Spain;
  - Oman to France, USA, Spain, South Korea, Japan and India;
  - Qatar to Belgium, USA, United Kingdom, Taiwan, Spain, South Korea, Japan, Italy and India;

- United Arab Emirates to India and Japan;
- Yemen to South Korea, USA;
- Abu Dhabi to Japan, Europe, USA, Spain;

- Atlantic Basin - Russia to India;
- Pacific Basin - Australia to India; and Malaysia to India

Most of the trade movement in the Indian Ocean involves the Middle East as the main supplier, led by Qatar and Oman. The emergence of India as the major importer in the Indian Ocean Rim contributes more transactions of LNG shipments across the Indian Ocean. Further, the development of LNG shipyards in India is another factor that has made the Indian Ocean the busiest route from 2005 onwards (this is based on the trade projection).

LNG Transportation: Safety Concerns

The rapid expansion of the LNG sector has brought another level of security, awareness and requirements. There is a need for close attention to be paid to the quality of vessels, terminal facilities and crews, for which it has become more difficult to assure. The expansion of the industry demands the importance of ensuring that LNG vessels and land-based facilities be equipped with greater security and safety aspects and staffed by qualified personnel with up-to-date training.

So far, LNG has been reported to have an excellent safety record at sea. For the past 40 years with more than 40,000 voyages, there has been no reported case of an LNG spillage by LNG tankers at sea. This is because LNG tankers have maintained a safe record of accomplishment and reputation for being well-built and well-maintained vessels compared to other liquid and gas tankers.

Despite the assurance by the industry that LNG is safe for transport by sea and storage in the tankers compared to other liquid fuels, it is not impossible that there could be a LNG spill and for an accident to occur. There have been a few studies on safety and security of LNG at sea and land-based facilities. Most of the studies have revealed that there is a possibility for a terrorist attack on an LNG carrier that could, under certain conditions, create an LNG explosion. Furthermore, breaching of an LNG tanker and spillage and dispersion due to ship collision and engine failure could lead to another environmental disaster.

A Report by Sandia National Laboratories revealed that the risk and hazards from an LNG spill is variable, depending on the size of the spill, environmental conditions and the site at which the spill occurs.[11] However, it also depends on the tonnage of the LNG tanker. A bigger LNG tanker may result in a bigger risk and hazard to the environment and to people. It was reported that LNG hazards may result from three of its properties - cryogenic temperatures, dispersion characteristics and flammability characteristics. The extremely cold LNG can cause injury and damage as explained earlier (California Energy Commission, 2006).

Safety Concerns
LNG carriers are specially designed with a double hull to provide optimum protection for the cargo and ship's crew in the event of collision or gas leakage. The liquid is loaded and stored into the ship's pre-chilled cargo tanks at atmospheric pressure in either double membrane containment systems or special three-quarter inch thick spherical tanks. The insulated tank will prevent leakage and helps to keep the LNG cold at approximately -163¡C. The containment system consists of a primary liquid barrier, a layer of insulation, a secondary liquid barrier and a second layer of insulation. The second layer of insulation will prevent leakage should there be any from the first layer of insulation. The safety aspects of the LNG tanker will keep the LNG cold at atmosphere temperature until it arrives at the regasification plant or port.

In its liquid state, natural gas is not explosive. LNG will only burn if it escapes into the atmosphere and mixes with air. If the LNG is spilled or leaks, it will immediately evaporate and form into a combustible vapour cloud, which if ignited can be quite dangerous. Once in the open air, the fire results in a visible cloud that can melt steel at a distance of 1,200 feet. The effects could be worse as the ignited fire can cause second-degree burns on exposed skin two miles away and produces a fire a mile wide from the site.[12] The fire at this scale would not be impossible to extinguish, as it will burn until the fuel is totally spent.

As described above, it raises a serious question over whether or not LNG is safe for international shipping. The potential effects of its risks and hazards are a concern to the industry players, local authorities and environmental activists. The industry has already made clear what impact there would be should there be any LNG explosion incident during its transshipment. Such incidents will not only cripple the LNG supply chain, but also effect the production of LNG, as it will have to slow down or to be stopped until the safety aspect is proven.

From 1965 until 2005, there have been four accidents involving LNG at ports, eight accidents during loading and unloading of LNG cargo and 13 accidents at sea. [13] Overall, most cases involved technical failure such as engine breakdown, valve leakage, overfilling collision and grounding. Out of 25 accidents, only one accident in 1996 led to six deaths and three injured at sea. The accident that occurred in October 1996 occurred when carbon dioxide (CO_2) was released accidentally from a fixed fire extinguishing system that was being tested by one surveyor and five shipyard technicians. However, there has been no spill or gas release from a storage tank due to any incident. It is important to note that an

LNG explosion will only occur if the spill forms into a vapour cloud when mixed with air. Only the right concentration of LNG vapour in the air will cause the overpressure of explosion. Besides, LNG vapour that is mixed with air is not explosive in an unconfined environment.

Until 2005, there has been no record of major accidents by LNG seaborne transport. For more than 40 years of LNG operations worldwide, there has been no significant release of LNG that was caused by breach or failure of a cargo tank, serious collision with other vessel, and massive attack during wartime - for example, the Iran-Iraq War in the 1980s (Michael Richardson, 2004). This has placed LNG tankers among the safest vessels on earth compared to other oil and gas vessels. Why it is so?

It is essential to note that the industry has acknowledged the potential of LNG as a dangerous cargo. Considering the possibility of the major impact of a LNG accident to the environment and human beings, the industry has already make sure that LNG transportation is equipped to the highest levels of safety and reliability.

Safety aspects
LNG carriers are fitted with sophisticated safety systems that consistently monitor the condition of the cargo tank and ship. All LNG carriers will have to comply with all relevant local and international regulatory requirements to enable the ship to operate at sea. The most prominent international conventions for LNG carriers to comply with are the International Gas Code (IGS) and International Ship and Port Facilities Security Code (ISPS Code) of the International Maritime Organization (IMO). In addition to that, it is compulsory for the LNG ship's officers and crews to undergo extensive training on how to operate the LNG ship. This is necessary to ensure that they have met the international standard.

The LNG ship is equipped with safety equipment to facilitate ship handling and cargo system handling from the liquefaction plant to the regasification plant. The ship handling safety features include global positioning and communications systems that enable the ship to monitor its position, speed and other vessels nearby. An automatic maritime distress system on the LNG ship is designed to transmit signals and an alarm should there be any onboard emergency that requires immediate external assistance.

The cargo safety system, on the other hand, includes an emergency shutdown system that can identify potential safety problems and automatically shut down the ship operations. This will prevent and limit the amount of LNG that can be released from the cargo tank. In addition, gas and fire-detection systems, nitrogen purging, double hulls and double containment tanks or leak pans are placed on board to help identify the risk of fire or leakage. These detection equipments were reportedly so sensitive that they could detect any leakage through a hole of a pinhead. Should there be any leakage or fire during its transshipment, the safety system will respond automatically.

It is different with other oil and liquid carriers since LNG ships use velocity meters to ensure speed during berthing. The automatic mooring line (on ship) will be connected to the onshore system, the instrument systems and the shore-ship LNG transfer system as one system during its mooring to the shore.[14] Should there be any leakage or accidental fender-bender, the connection between both safety system on ship and shore will automatically shut down the entire system and discharge operations can be shut down remotely.

The safety system as described above is designed to secure and prevent the probability of LNG incidents and the potential consequences. As mentioned earlier, it is very important to note that in more than 40 years history of LNG shipping, there have been no LNG cargo failures, leakage that has led to vapour clouds and collision that has resulted in a major disaster.

Conclusion

There are a number of factors that have stimulated growing demand in the LNG industry. The flexibility of the gas to the electrical power sector has increased the interest worldwide in replacing the use of oil and coal in the energy sector. Notably, the gas comprises an environmental-friendly source compared to oil and coal.

Lower costs in liquefaction and shipping have contributed further to the development of the LNG industry. Major oil companies - ExxonMobil, Shell and BP - have invested a huge amount of capital in developing a larger LNG storage, replacing the small and uncompetitive storage. The development in engineering techniques and technology provide an efficient way in managing the stranded resources into full production. In addition, high prices of natural gas in the market have increased the desire of LNG suppliers/producers to monopolise their gas reserves.

Moreover, there has been a radical development in LNG carriers. The capacity of the LNG vessels has continuously increased whereby they provide more space for cargo utilisation. It is a fact that the LNG carriers can trade longer compared to ordinary vessels in the same age range. This is due to the quality and maintenance of these vessels. In addition, the complexity of the vessels minimise the adverse effects of any mishap during its transshipment from liquefaction plant to regasification plant. The safety record of LNG carriers in the past 40 years has reassured the industry players that LNG provides less risk in terms of environment and human hazard and minimum loss projection in trade.

It should be noted that there has indeed been an improvement in operational safety in the past 40 years of operation. Effective training for the personnel, in detailed procedures for loading and unloading the cargo to the storage plant, unique regulations by each state and security and prevention measures are already in place today. Undoubtedly, the industry has taken all necessary steps to secure the safety and security of the LNG industry. Driven by these factors, there is a huge potential for natural gas in the energy industry. With natural gas liberalisation, there is no doubt that the industry will reduce the reliance on the oil and coal industry over at least the next 20 years.

Endnotes

1) Relating to or producing low temperatures

2) Trillion cubic feet

3) Eastern Europe/ Former Soviet Union

4) LNG Development Question Becomes 'How' rather than 'Whether'. [WWW]<URL:http://www.cera.com/news/details/print/1,2317,7735,00.htm >[Accessed 12 April 2006]

5) The impact of Growth in Korea on Asia LNG Supply and Demand - A Supplier's Perspectives.[WWW]<URL:http://www.woodside.com.au/NR/rdonlyres/ ehamgssrk7455alyhta7bla6rr4lokyd3edqagvna2247dmeh22d374favolnntrzszcqtr7 wf5eje/Resource8.pdf >[Accessed 5 November 2005]

6) Trillion cubic metres

7) The Global Liquefied Natural Gas Market: Status and Outlook. [WWW]<URL:http://www.eia.doe.gov/oiaf/analysispaper/global/exporters.html> [Accessed 13 May 2005]

8) Brinded, M. 2003. Shared Trust- The Key to Secure LNG Supplies.[WWW]<URL: http://www.shell.com>[Accessed 5 November 2005]

9) LNG Newbuilding in a nutshell. Gard News. November 2005 - January 2006. 7-8.

10) Panorama 2005. Outlook for the World Gas Market. [WWW]<URL: http://www.ifp.fr/IFP/en/files/cinto/IFP >[Accessed 18 May 2006]

11) Sandia National Laboratories. 2004. *Sandia Report on Guidance on Risk Analysis and Safety Implications of a large Liquefied Natural Gas (LNG) Spill over Water.* California: Sandia National Laboratories.

12) Raines, B & Finch, B. 2003. Study talks of possible LNG disaster as result of accident. [WWW]<URL:http://www.wildcalifornia.org/cgi-files/0/pdfs/ 1076793906_Humboldt_Bay_LNG_Mob_Reg_Study_Possible_Disaster.pdf -. > [Accessed 19 April 2006]

13) Accidents and Problems Involving LNG Carriers.
[WWW]<URL:http://www.coltoncompany.com/shipbldg/worldbldg/gas/lngcaccid
ents.htm >[Accessed 17 May 2006]

14) Fross, M.M. 2004 The Role of LNG in *North American Natural Gas Supply and Demand*.
[WWW]<URL:http://www.beg.utexas.edu/energycon/lng/documents/CEE_Role_of
_LNG_in_Nat_Gas_Supply_Demand_Final.pdf >[Accessed 4 November 2005]

References

Pitblado, R.M. et al. 2004. Consequences of LNG Marine Incidents. [WWW]<URL:http:www.piersystem.com/posted/795/FINAL_CCPS_Paper_Cons equences_of_LNG_Marine_Incidents_Final.59347.pdf > Accessed 12 June 2005

Sandia National Laboratories. 2004. *Sandia Report on Guidance on Risk Analysis and Safety Implications of a large Liquefied Natural Gas (LNG) Spill over Water.* California:Sandia National Laboratories.

Richardson, M. 2004. *A Time Bomb for Global Trade & Maritime-Related Terrorism in an Age of Weapon of Mass Destruction.* Pasir Panjang: Institute of Southeast Asian Studies Singapore.

International Energy Outlook 2004-Natural Gas. [WWW]<URL:http:// www.eia.doe.gov > Accessed 13 June 2005

Institute for Energy, Law and Entreprise. Introduction to LNG. *An Overview on Liquefied Natural Gas (LNG), Its Property, the LNG Industry, Safety Considerations.* [WWW]<URL:http://www.beg.utexas.edu/energyecon/lng/documents/CEE_INTR ODUCTION_TO_LNG_FINAL.pdf > Accessed 13 June 2005

World LNG Industry Review. . [WWW]<URL:http://www.ilnga.org > Accessed 13 June 2005

Worldwide Natural Gas Supply and Demand and the Outlook for Global LNG Trade. [WWW]<URL :http://www.tonto.eia.doe.gov > Accessed 13 June 2005

The Indian Ocean Region: An Australian Perspective. [WWW]<URL:http://www.dfat.gov.au/archive/speeches_old/minfor'geind.html > Accessed 13 June 2005

IAMU LNG Round Table on Joint Statement on LNG Ship's Officer Competency Standards by The International Association of Maritime Universities (IAMU) and The Society of International Gas Tanker and Terminal Operators (SIGTTO), 1st March 2005. [WWW]<URL: http://www.iamu-edu.org/workinggroups/lng.php > Accessed 19 May 2005

Qatar Country Analysis Brief. [WWW]<URL:http://www.eia.doe.gov/emeu/cabs/qatar.html > Accessed 8 May 2005

The Impact of Growth in Korea on Asia LNG Supply and Demand- A Supplier's Perspective. [WWW]<URL:http:// www.woodsde.com.au> Accessed 19 May 2005

Natural Gas Changes in Indonesia.
[WWW]<URL:http://www.usembassyjakarta.org/econ/natural_gas2003html > Accessed 19 May 2005

BP 2004, Statistical Review of World Energy. [WWW]<URL: http://www.bp.com> Accessed 19 May 2005

The Global Liquefied Natural Gas Market: Status and Outlook. [WWW]<URL:http:www.eia.doe.gov> Accessed 19 May 2005

Outlook for the World Gas Market. [WWW]<URL: http://www.ifp.fr/IFP/en/files/ cinfo/IFP > Accessed 19 May 2005

World LNG Industry Review. [WWW]<URL: http://www.ilnga.org > Accessed 20 May 2005

Presentation paper on Shared trust- the key to secure LNG Supplies by Malcolm Brinded, Group Managing, Royal Dutch/Shell Group, CEO, Shell Gas and Power. [WWW]<URL: http://www.shell.com-static-media-en-downloads-speeches-mb_lng_supplies_17122003.pdf.url > Accessed 20 May 2005

Annual Energy Outlook 2005. [WWW]<URL: http://www.eia.doe.gov > Accessed 27 May 2005

LNG the Growing Alternative Emergence of a US Market & Role of Qatar as an International LNG Hub. [WWW]<URL: http://www.csis.org/energy/040316_ verrastro.pdf > Accessed 27 May 2005

Liquefaction natural gas transportation: in the pipeline. [WWW]<URL: http://www.teriin.org/energy/lng.htm > Accessed 27 May 2005

Storvik, C. (ed). 2006. LNG Newbuilding in a Nutshell. Gardnews. November 2005-January 2006. pp. 6-8.

Death and Injury Incidence Reports Associated with Carbon Dioxide Total Flooding Fire Extinguishing Systems. [WWW]<URL: http://epa.gov/ozone/snap/fire/co2/appendixa.pdf > Accessed 28 April 2006

Buckland, A. (ed). LNG Market. Drewry Shipping Insight. April 2006.pg 34

Accidents or Problems Involving LNG Carriers. [WWW]<URL: http://www.coltoncompany.com/shipbldg/worldsbldg/gas/lngcaccidents.htm > Accessed 28 April 2006

Havens, J. 2003.*Terrorism. Ready to Blow?* Bulletin of Atomic Scientists. July/August 2003. Volume 59, No.4, pp.16-18.

Noor Apandi Osnin. 2004. *Bulk Shipping Market Cycle in the period after World War II- the LNG Trade*. Kuala Lumpur: Maritime Institute of Malaysia.

SECURITY THREATS IN THE INDIAN OCEAN

Chapter 10

Assessing Insecurity in the Indian Ocean Region

Aparajita Biswas

Introduction

The late 20th century has seen a profound change in the concept of security. A major feature of the change was the conclusion of the Cold War, the end of major conflicts in the region, and the dismantlement of apartheid in South Africa, which paved the way for the current period of democratisation, regional cooperation and relative peace. However, despite these changes in international and regional relations, security threats have not been totally eliminated, but simply re-manifested. Non-traditional security threats like environmental degradation, mass population movements, small arms proliferation and drug smuggling, to name a few, have assumed increasing prominence.

This chapter is intended to address the issues of small arms and drug trafficking as well as the symbiotic relationship that exists between narco-traders and arms dealers. These issues will be examined by means of a regional geographical analysis with the Indian Ocean Region (IOR) being the focal point. As the IOR is vast, I have preferred to be selective. I have concentrated the investigation on a few South Asian and Southern African Countries in the IOR, where the situation of small arms and drug trafficking is most alarming. The chapter will also interrogate the possibilities of cooperation between regional organisations like the South Asian Association for Regional Co-operation (SAARC) and the Southern African Development Community (SADC), to counter this illicit network in the Indian Ocean Region.

The Trans-National Nature of (In)Security

The end of Cold War was followed by a rise in trans-national security issues. Many countries in the Asia-Pacific and the Indian Ocean regions are slowly recognising that trans-national security issues are emerging as their top security challenges, and pose even more long-term threats to state and regional security than inter-state conflict. At their most basic level, trans-national security issues can be defined as non-military threats that cross borders and either threaten the political and social integrity of any state or the health of its inhabitants. This can be broadly termed as low intensity conflict. Unlike conventional and nuclear threats, which are resorted to by states, low intensity conflict is the instrument employed by state-sponsored groups, as well by non-state actors - such as criminal gangs or terrorist groups - who hardly care for international laws and standards. Their causes are multifarious and are not easily ascertainable. It is a long-term problem characterised by the dilemma of finding the ideal solution - should it be a single policy change or by the introduction of an international law and convention (Gallaghar, 1992).

The unrestrained spread and the associated illicit trafficking of small arms and light weapons is not a new phenomenon in itself, but has attained a new dimension with the end of the Cold War. The funding link between terrorist groups and narcotics trafficking is well-known. The nexus between the two led to the term narco-terrorism. It may be mentioned that the term narco-terrorism was coined by President Belaunde Terry of Peru in the year 1983 when describing terrorist types of attacks against his state's anti-narcotics police. Now a subject of definitional controversy, narco-terrorism is understood to mean the attempts of narcotics traffickers to influence the policy of the government by the systematic threat or use of violence (G. Davidson (Tim) Smith, 1991).

On a regional and national level, drugs and small arms together can be destabilising influences. The small arms trafficking intimidates the state in several ways. First, it can undermine democratic institutions as governments seek to control the threat. Their attempts often result in draconian measures, such as military takeover of the civilian government. Added to this, small arms nurture domestic criminal groups that erode or confront the power of the state. In fact, the proliferation of small arms around the world partly reflects the "shift of armed conflict progressively from the regular to irregular". It is been institutionalised to such an extent that it has created a new kaleidoscope wherein neither the old rules nor new weapons apply. The developing world is most affected by it and yet, this threat has not been given the systematic academic attention that it deserves.

The South and Southeast Asian regions are particularly vulnerable to the scourge of small arms trafficking. This is due to various factors. The Afghanistan-Pakistan region arguably contains the world's largest concentration of the illicit weapons, a situation made more volatile by the fact that it is also a centre for terrorist and extremist ideology. Similarly, Asia has two of the largest opium-producing countries in the world - Myanmar and Afghanistan - and it is well-known that narcotics trafficking and its collateral violence, depend extensively on the availability of small arms.

In South Asia, the fallout of religious fundamentalism pursued by Pakistan as an instrument of regional policy and the subsequent post-Najibullah events in Afghanistan also saw the spread of several negative patterns for deeper ramifications for South Asian security. The large-scale weaponisation in Afghanistan, Baluchistan and the North West Frontier Province of Pakistan and the intertwined relationship of weapons production and the narcotics trade, lumped together as the 'Kalashnikov culture', has set off a fundamentalist drive into Kashmir, Tajikistan, Chechnya, Bosnia and elsewhere. India's vulnerability increased from the growing narcotics trade across borders, bringing with it a host of social and political problems. The harmful effects of armed conflicts, international terrorism, drug smuggling and the illicit trafficking in small arms has challenged world security and stability. National borders did not confine those threats and no individual country could cope up with those challenges. It should be confronted at the regional, national and international levels.

The point worth noting is that the Cold War period left behind a legacy of weaponry that could undermine the progress made to foster peace and democracy in the developing countries. During the Cold War, the United States and the erstwhile Soviet Union resorted to arming rebel groups throughout the world. The irony is that today many of these illicit transfers of arms by these two countries have come back to haunt them. For example, the USA is now engaged in a fight against Colombian drug lords who, ironically enough, are sometimes armed with the very weapons slipped to Central America by the USA to fight communism in the 1980s.

India is caught between the three largest heroin and opium producers in the world, Afghanistan, Pakistan and Myanmar, resulting in conflicts on her borders adjoining major drug production/transporting areas. Sri Lanka too suffers tremendously from the serious proliferation of small arms. The Sri Lankan militants have penetrated deeply into the drug world, to augment their armed strength against the Sri Lanka Army. Added to this is the fact that the notorious drug producing and illicit arms trading areas of the 'Golden Crescent' and the 'Golden Triangle' lie within the geographical propinquity of the IOR. This geographical association is further reinforced by the link between narcotics and arms, with the sea routes of the Arabian Sea and the Bay of Bengal providing ideal waterways for the supply of both arms and narcotics (Rao, 2003).

In fact, one critical problem arising out of the confluence of two major narco-producing and trading regions - the Golden Crescent and the Golden Triangle - is the cementing of a diabolic relationship between insurgent groups, arms dealers and narco-terrorists. Such a relationship is quite common with the groups operating in the region. Although not all the insurgent groups engage in narco-production or narco-trafficking, it has nevertheless been found that all of them have regularly taxed and extorted money from the traffickers, while providing protection to the latter for conducting trafficking in drugs. There are several critical implications to this.

Firstly, the trans-national narco-networks, now backed by armed insurgents, make an anti-narco-production or narco-trafficking drive immensely difficult. And taking into consideration the geographical and topographical conditions in which the insurgents and the traffickers operate, there is now all the more reason to believe that the nationally-organised military or coercive solutions may not be the correct way of overcoming the menace of narco-terrorism.

Secondly, weapons, particularly small arms in the hands of both the insurgents and traffickers, become more rampant, to the point of threatening the law and order situation in the vicinity. A large portion of the money received from taxing and extorting the narco-traffickers goes towards the purchase of small, at times sophisticated, arms for the insurgents (Rubin, 1991).

In Africa too, the same scenario persists. The end of the Cold War did not see Africa become a more peaceful place. The dissolution of the bi-polar world order was also accompanied in some cases by the collapse of individual states. Coupled with it there was a gradual erosion of interests among the major world powers towards Africa. In the void left by the crumbling state and the international community have stepped external non-state actors, sometimes as proxies or sometimes as independent agents, who have influenced events and fuelled conflicts. Millions of small arms - lightweight, portable and devastatingly - effective even in the hands of the young or poorly-trained users were shipped to Africa during the Cold War, to equip anti-colonial fighters, newly-independent states and superpower proxy forces alike. Even after the Cold War ended, a flood of new small arms entered Africa as arms manufacturers put millions of surplus Cold War era weapons into the international arms market, at cut rate prices. In July 2001, the US Government estimated that small arms have fuelled conflicts in 22 African countries, leading to a loss of 7.8 million lives. Indeed, in Africa, guns are more than weapons of choice - they are weapons of mass destruction.

In the Southern African region the situation is most volatile. The region has recently emerged from a period of protracted violent conflict characterised by a series of national liberation insurgencies and civil war. Currently, it is undergoing a term of relatively peaceful democratic consolidation. Despite these developments, uncertainty in terms of potential threats to the region and localised conflicts, which have been sustained by massive socio-economic disparities and mass movements of refugees, present a major obstacle to achieving peace and human development. Local conflicts have been exacerbated and even intensified by the availability of small arms. The result was that, years later these durable killing machines have fallen into the hands of insurgents, local militants and criminal organisations. This has left ordinary people vulnerable to violence arising out of ineffective policing and simmering civil conflicts (Reyneke, 2000).

This chapter would like to examine the continuing trade in light weapons, and its nexus with the narcotics movement, where it exists in the Indian Ocean Region. The argument has been made that the weapons and narcotics movements respect no frontiers, their passage is accelerated by underdevelopment, ignorance and economic decay, which destabilise the entire region. Arms spill over from conflict to conflict, and they serve the greed and need of criminals who recognise no borders. It is not possible for a single state to handle and control the situation, underlining the absolute necessity of evolving a regional approach to solutions, moving in parallel with a global awareness of the problem and a determination to deal with it.

In broadly classifying the countries in the Indian Ocean Region, it is certain that some countries are primarily weapons repositories, while others are mainly drug producers. In the African littoral states of South Africa and Mozambique, small arms have played a major role in exacerbating crime and armed violence. Somalia offers a textbook case of the dangers of light weapons proliferation and irresponsible arming of unstable regimes by the former powers. However, in all of these states, smuggling of drugs and weapons are a follow-up symptom rather than the disease itself.

However, this classification does not hold good in the case of countries like Pakistan and Afghanistan. In both these countries, statistics show that the availability of weapons and drug cultivation has reached an all-time high, and the growth of narcotics cultivation is synonymous with the rise of the conflict itself. These commodities flow out into the IOR through the ports of South Asia, on their way to the lucrative markets of Europe and USA. In short, the Indian Ocean sea lanes of communication function as trafficking routes for the regional and global trans-nationalisation of insecurity.

The Circulation of Small Arms

The terms 'small arms' and 'light weapons' - often used coterminously - have come into common use in recent years. The definition of light arms and light weapons used by the Small Arms Survey covers both military-style weapons and commercial firearms (hand guns and long guns). Small guns include weapons like revolvers and self-loading pistols, rifles and carbines, assault rifles, sub-machine guns and light machine guns. Light weapons are heavy machine guns, hand-held under barrel and mounted grenade launchers, portable and anti-aircraft guns, recoilless rifles, portable launchers of anti-tank and anti-aircraft missile system and mortars of less than 100-meter calibre (UN, 1997).

These weapons have become easier to obtain, easier to conceal and to smuggle across borders, and easier to use and maintain to the point that it has resulted in the phenomenon of the child soldier or child combatant, turning innocent children, many of them as young as seven, into ruthless warriors and killers. These weapons are increasingly lethal, with the most recent models capable of firing as many as 700 rounds a minute. According to Ado Vaher of the UNICEF, small arms and light weapons are now responsible for no less than 90% of war casualties. Since the 1990s, more than 3 million people had been killed by small arms and light weapons. Among them, 80% are of innocent civilians with children accounting for 25% of all casualties. This means that the deaths of 750,000 children were directly caused by small arms and light weapons (UN, 2004).

Worldwide, AK47 (or Kalashnikov) rifles are proving to be extremely popular small weapons. Within the AK series of assault rifles alone, an estimated 78 countries are using them, while more than 14 countries have manufactured between 35 and 40 million AK assault rifles in the period between 1945 and 1990. The proliferation of these rifles is mainly due to their ease of production, their excellent performance under adverse conditions, and, most importantly, their ease of use and disassembly. AK 47 rifles, the nodal point of the AK family, has achieved unparalleled levels of availability in the global arms market. These rifles are user-friendly and can be easily handled even by child-soldiers. Such is the ubiquity of the AK rifle that it has become a symbol of resistance and used by guerrilla movements worldwide, including the Afghanistan Mujahideen, the African National Congress, the Irish Republican Army, UNITA, and the Kosovo Liberation Army. An indication of the extent of proliferation can be measured from the fact that, in some parts of Africa, AK47 rifles can be purchased for as little as US$6, or traded for a chicken or a sack of grain.

The most classic situation is in Afghanistan, where the US sent billions of dollars of military aid to radical Islamic fundamentalists in 1980 to fight the Soviet-backed regime of Babrak Kamal. This country received large volumes of weapons in the 1980s, and is today a major source of small arms and light weapons supply in South and Central Asia. During the Afghan War between 1979 and 1989, the erstwhile USSR, on the one hand, and groups of states on the other, pumped in millions of small arms into the region. According to sources, the USSR incurred an intervention cost of around US$3billion a year up to 1992.

In Afghanistan, the billions of dollars that the USSR spent on the regime were, in terms of quantity, far less than the billions that are spent in buying tons of assault rifles, RPGs, and mortars, to sustain resistance movements. Other sources have put the cost of the Afghan conflict at US$5billion, which was about fifty times the yearly aid that was given by the USSR to the Afghan regime to 1989. In addition, ten million anti-personal mines were laid by the then Soviet troops during their 10-year occupation of Afghanistan. The mining operation continued till their withdrawal in 1989, covering nearly 13 million square meters of land - from rural landscape to urban thoroughfares. Only 43,000 mines have been cleared so far. Most of the agricultural land in Afghanistan is infested with mines and the rest is under poppy cultivation. The erstwhile Soviet Union is known to have spent as much as US$20 billion on arms supplies during their Afghan occupation. Some of these arms have undoubtedly fallen into the hands of Afghan warlords, the real beneficiaries of the largesse from the Inter Services Intelligence (ISI) of Pakistan (Klare, 1995).

It is a common knowledge that during the Afghan-Soviet war in the 1980s, both the CIA and the ISI played a direct role in funnelling weapons and money to the so-called 'freedom fighters' - the Afghan Mujahideen. The two agencies were also at least indirectly involved in the nascent, local trade in narcotics. The Mujadeen received most of the funds supplied by the CIA (and matched by Saudi Arabia) through the ISI, where national logistic cell trucks delivered weapons to Afghanistan and brought opium to Pakistan. The CIA channelled at least US$2billion in weapons aid, or an estimated 80% of the agency's covert aid budget to the Mujahideen in Afghanistan, in its fight against the then Soviet forces.

The involvement of the ISI in both arms and drugs trades raises further questions about the adequacy of a category such as 'narco-terrorism', and would require an assessment of how such different actors as resistance guerrillas, intelligence agencies and terrorist organizations use the drug economy. It has been reported that 340 Stinger missiles were fired during the Afghan war, bringing down 269 aircraft, a 70% success scored by marksmen who were not even fully trained. The paradox is that, the USA is now worried that the deadly Stinger - supplied by it - might fall into the hands of the terrorists. According to sources, Stingers were resold to countries like Libya, Iraq, and North Korea.

It was partly as a result of this practice of siphoning off a portion of the arms in transit, that Pakistan has become a major source of small arms supplies in South Asia. This covers both black market arms and arms supplied covertly to insurgent groups in the region. While there are some ideological incentives for transferring arms, some tribal groups (for example, the Baluchistan region in Afghanistan) view these transactions mostly from a financial point of view (Agha, 1995).

Under the US Reagan Administration, it was believed that the arms aid amount in the Afghan conflict jumped from US$80 million in 1983 to US$120 million in 1984. These figures climbed to US$470 million in 1986 and US$630 million in 1987, and covered arms ranging from M-16 and UZIS to Kalashnikovs of different makes, including Russian, Chinese and Eastern European manufactures (Klare, 1995, p. 13). Total aid from all sides has been put by other sources at around US$8.7 billion (1986-1990), far exceeding the legal weapons imports of most countries during the same period.

There is clear evidence that the instances of violence in India and Pakistan are a direct outcome of the irresponsible arming of Afghan Mujahideen by the United States, with considerable assistance from Saudi Arabia and partly from Britain. In this context, it is to be mentioned that the flow of weapons into the region has clearly played a major part in the erosion of law and order over the past decade (Smith, 1993). Large numbers of pipeline weapons have made their way into the hands of Sikh and Kashmiri militants. The militants obtained weapons either directly from the ISI of Pakistan or from the arms bazaars situated in the North West Frontier Province.

In India, while many factors contributed to the growing strength and resolve of Sikh fighters during the mid-1980s and through the early 1990s, increased access to vast quantities of more advanced weapons, however, has allowed them to consolidate power through force. At the same time, the acquisition of large numbers of these weapons contributed to the dramatic increase in both the frequency and severity of abuses inflicted on the unarmed civilians. It is clear that there is a correlation between the number of civilians killed and the use of automatic rifles; as the use of Kalashnikov's increased, so did the number of civilian killings (Singh, 1995).

The other disquieting factor is that India's rebel groups also receive weapons originating in Pakistan. In the northeastern part of India, there are three insurgency groups - the United Liberation Front of Assam (ULFA), the National Socialists Council of Nagaland (NSCN) and the National Democratic Front of Bodoland (NDFB). These groups are in direct contact with Pakistan's ISI and the Afghan Mujahideen. Not only do they get training in those countries but also receive weapons through a pipeline originating in Pakistan, and the Southeast Asian pipeline running through Thailand, Malaysia and Singapore (The Hindu, 2000).

It may be mentioned here that the Andaman and Nicobar Islands in the Indian Ocean are located in the most insecure region. The waters around the islands are routes for the Chinese gun-running to Bangladesh, from where the arms reach northeast India. It was mainly for this reason that the Indian government upgraded the security structure of the islands and the surrounding areas in 1998. It had established a High Service Command to combat the smuggling of narcotics and arms, and to keep an eye on illegal shipping and other maritime activities in the region. As regards strategic concerns, the Andaman and Nicobar Islands provide the key to the eventual success of the much talked about 'Look East Policy' of India that was enunciated by the then Prime Minister of India, Narasimha Rao, in the early 1990s. It is the close proximity of the Andaman and Nicobar Islands with southeast Asia that makes India as much a part of this region as South Asia.

According to Indian Intelligence Reports, the chief patron of the ULFA is Pakistan. The ULFA has for long maintained close linkages with the Pakistan ISI, which procured several passports for its chief, Paresh Barua, and the other ULFA cadres. The top ULFA leaders are in close touch with certain officers of the Pakistan High Commission in Bangladesh, who have arranged for their passports in various names to travel to Karachi, on the way to various training camps run by the ISI and its affiliates. At least 300 ULFA cadres were trained in Pakistan in the use of rocket launchers, explosives and assault weapons. There are several madrassas and mosques sponsored by the ISI in the Sylhet and Cox Bazaar areas that are being used to hoard and transfer arms procured by the ULFA from Thailand and Myanmar (Subir, 1996). The ISI largesse enabled the ULFA to buy arms in Cambodia, paying for them in hard currency routed through Nepal. Their allegiance to Pakistan is so intense that they openly supported Pakistan during the Indo-Pakistan conflict in Kargil.

Furthermore, according to Indian intelligence sources, the ULFA is receiving direct support from Bangladesh and China. It has established close contact with the Chinese army and through them, arms have been transported from China to ports in Bangladesh on merchant ships of various countries, including North Korea. The ULFA camps in Bangladesh have been functioning since 1989. Commencing initially with using Bangladesh as a safe haven and training location, the ULFA gradually expanded its network to include operational control of activities and the receipt and shipment of arms in transit before they entered India. The Muslim United Liberation Tigers of Assam (MULTA) and the Muslim United Liberation Front of Assam (MULFA) are the chief suppliers of arms for the ULFA through Bangladesh. This group has reportedly set up bases in the hill region of Meghalaya, to coordinate the transit of arms coming through Bangladesh.

Moreover, the cooperation between various terrorist organizations in India's northeast and foreign groups was formalised with the formation of the Indo-Burmese Revolutionary Front (IBRF) in 1989. The IBRF was made up initially of the NSCN-K, ULFA, NDFB, Kuki National Front and Chin National Front of

Myanmar. The ULFA and the NFDB have also developed contacts with the LTTE (Liberation Tamil Tigers of Elam), the rebel group in Kashmir, the Kachins in Myanmar and the Rough in Cambodia (Subir, 1996). The NSCN, which is heavily armed, constantly provokes insurgencies in the neighbouring states of India, such as Mizoram, Manipur, Tripura and Assam. It also receives active political and material support from India's hostile neighbours.

Another country which suffers tremendously because of the serious proliferation of small arms is Sri Lanka. The Tamil militants have purchased explosives and weapons from a wide variety of sources such as North Korea to Myanmar and Ukraine and from middlemen operating from Europe to Asia and the Middle East. The accumulation of small arms in Sri Lanka is driven by the rebel secessionist movement led by the LTTE fighting for independence from Sri Lanka in the northern part of the country. Within two decades, the LTTE had emerged as one of the world's most feared guerrilla groups.

It is a well-known fact that prior to 1987, the LTTE guerrillas were trained in several camps in India and were permitted to run their own camps in the country. During the latter part of the 1970s and the first seven years of the 1980s, New Delhi played a key role in supporting the militant Tamil struggle in Sri-Lanka, backing both the LTTE and several other militant groups. This support was a product of wider ideological and geopolitical concerns. However, after the arms flow from India dried up following the Indo-Sri Lanka Peace Accord, the LTTE has successfully replicated that network internationally. It is to be mentioned here that the LTTE refused to accept the peace accord which aimed to bring about a negotiated settlement to the insurgency in Sri Lanka by providing for a general ceasefire and the devolution of local powers of governance for an autonomous northeastern Tamil province. The Indian Peace Keeping Force was despatched to Sri Lanka to oversee guerrilla disarmament and provide security to local administrative council elections. The LTTE regarded the Indian policy reversal as an unforgivable act of treachery, which then influenced its decision to assassinate India's then Prime Minister, Rajiv Gandhi.

The LTTE arms network is headed by Kumaran Pathmanathan, aka 'KP'. At the heart of KP's operations is a highly secretive shipping network. The LTTE started building its maritime network with the help of a Bombay Shipping magnate in the middle of the 1980s. It is believed that the LTTE has around 10 freighters, which are equipped with sophisticated radar and Inmarsat Communication technology. The vessels mostly travel under Panamanian, Honduran or Liberian flags (famously known as Pan-Ho-Lib).

The communication hub of the LTTE weapons procurement network are southeast and east Asian locations such as Singapore and Hong Kong, which are not only strategically situated on key shipping lanes, but also have highly developed banking structures. These two 'city states' orchestrate cells located in Thailand, Pakistan and Burma, effectively plugging the LTTE into the booming arms bazaars of southeast and southwest Asia.

More distinctively, the international arms procurement activities of the LTTE are carried out from 5 main geographical locations. These are; 1) Northeast and Southeast Asia, focusing particularly on China, North Korea, Cambodia, Thailand, Hong Kong, Vietnam and Burma; 2) Southwest Asia, focusing particularly on Afghanistan and Pakistan (through the so-called Afghan pipeline); 3) The former Soviet Union, focusing particularly on Ukraine; 4) Southeastern Europe and the Middle East, focusing particularly on Lebanon, Cyprus, Greece, Bulgaria and Turkey, and, 5) Africa, focusing particularly on Nigeria, Zimbabwe and South Africa (Chalk, 1996).

The LTTE has also established well-connected traffic routes that are used to transport armaments back to Sri Lanka. Weapons from China, North Korea and Hong Kong are trafficked across the South China Sea, through the Malacca and Singapore Straits to the Bay of Bengal and then to Sri Lanka. The LTTE established a permanent naval base in Twante, an island off Myanmar, while Phuket in Thailand has been an important back-up base for them. Arms from Cambodia, Vietnam and Burma transit through Thailand before being loaded into vessels at the southern port of Ranong for the trip across the Bay of Bengal. Weapons from Eastern Europe, Ukraine and the Middle East pass through the Suez Canal, around the Horn of Africa and then on to Sri Lanka. Finally, arms from Africa tend to be smuggled back to the LTTE jungle strongholds, either around the Cape of Good Hope from ports in Liberia, Nigeria and Angola via Madagascar from the Mozambican coastal town of Beira (Chalk, 1996).

However, what is ironic is that the LTTE has obtained arms from the Sri Lankan Government itself. Sources in India alleged that thousands of arms went to the LTTE in 1989, ostensibly for them to fight against the Indian Peacekeeping Forces in Sri Lanka. There is sufficient evidence to show that the purchase of these weapons is primarily funded by the Tamil diaspora based in Switzerland, Canada, Australia, USA, UK and the Scandinavian countries. Most of these Tamil people are relied on by the LTTE, to facilitate their integration into the host society for forged identity papers, jobs and housing. It is estimated that the LTTE raises about US$2 million from this Tamil diaspora on a monthly basis, a quarter of it coming from Canadian Tamil expatriates. However, the collection of funds from Tamil expatriate sources is insignificant compared to the money obtained from narcotics. For instance, the current cost of a single hit of heroin (less than a gram) is around 10 British Pounds, while the wholesale price of a kg sold in New York is estimated to be at least US$250,000.

The other Indian Ocean Region areas which suffer from massive availability of small arms and light weapons in recent years are South Africa and the Southern African region. In Mozambique, arms dealers can purchase AK 47 assault rifles, together with a couple of clips of ammunition, for as little as US$14, or simply exchange a bag of maize for one. In Uganda, at the same time, an AK 47 can be purchased on the illicit market for US$10, the equivalent of the cost of a chicken. It has been estimated that as many as 6 million AK47s remain at large in the country (Fung, 2000).

During the apartheid years, the Government of South Africa began supplying tons of arms and ammunition to its domestic and regional allies for the defence of white minority rule. On the other hand, the opposing forces, the anti-apartheid groups, also smuggled an estimated 30 tons of arms and ammunition into South Africa, besides stockpiling arms at their base camps in surrounding countries. At the end of 1995, there were more than 4 million licensed weapons owned by two million people.

Figures on the number of illegal small arms in South Africa vary between 400,000 to 4 million. However, the extent of the problem can be gauged from the fact that the major portion of the arms has entered the illegal market. In 1999, 4% of all murders and 76% of all robberies in South Africa were committed with the use of firearms (Ministry of Safety and Security, 2000, p. 168). The presence of so many weapons outside government control has undermined law enforcement efforts, contributed to crime and public insecurity, hampered economic growth and caused tragic and avoidable deaths.

The Drug Network

The main drug of interest to trans-national criminal groups in the IOR is heroin, which is derived from the opium poppy. The opium poppy is cultivated in two main areas referred to earlier - the 'Golden Triangle' and the 'Golden Crescent'. The Golden Triangle refers to a roughly triangular zone in the highlands of Southeast Asia that overlaps Burma, Thailand and Laos. Although divided among the three countries, it is a region that shares significant attributes - opium production, remote upland terrain, mountain minority populations, extreme ethnic diversity, mass Christian conversion, and a long history of insurgency. The area of the 'Golden Crescent' consists of Afghanistan, Pakistan and Iran. It has emerged as the leading heroin-producing region in the world.

In the early 1990s, heroin became a leading illicit narcotic, and something of a world 'drug'. Almost 80% of heroin in Europe and 20% of heroin in USA comes from this region. Simultaneously, the end of the Afghan war and the repatriation of the refugees led to an expanded local heroin production. Heroin production received a boost because of the increasing potential for opium farming by Central Asian countries like Uzbekistan and Kazakhstan, through their ethnic links with Afghanistan. Drug trafficking (and the laundering of drug money) accounts for what is by far the most important category of illicit trade flows in the Indian Ocean Region.

The President of Germany's Federal Office of Criminal Investigation disclosed at a conference in Berlin in 2000, that the 'Golden Crescent' of Afghanistan, Iran and Pakistan had replaced the 'Golden Triangle' of Myanmar, Thailand and Laos as a major source of illegal drugs for Eastern Europe and the 'silk route'. Afghanistan is the largest opium-producing country in the region, producing 4600 tonnes of opium on an estimated 91,000 hectares of land (total world production of opium is 6000 tonnes). It accounted for 70% of the world's illicit opium production, which is converted to heroin. Although the cultivation of the opium poppy was officially banned in 1957, the ban was ineffective because of weak administration, the inaccessibility of the production areas and the domination of local tribal forces.

Pakistan alone has the second largest share of heroin production amongst the three drug-producing countries in the region. In South Asia, the drugs trade out of Myanmar, Afghanistan and Pakistan is estimated to be worth US $200billion.

In fact, the Taliban regime of Afghanistan had covertly encouraged opium production and collected a 20% tax from opium dealers and transporters, which went straight to the war chest. Over 80% of opium is grown in Afghanistan's northern province of Nangarhar and in the Helmand valley in southern Afghanistan. The ethnic bonds between the Baluchi tribesmen on both sides of the Afghan-Iran border provide the trans-national link which safeguards the traffic. In the backdrop of the disastrous economic conditions in Afghanistan, the drug trade seems to be a very lucrative business.

The major cause for concern was that, between 1979 and 1989, for the first time in the history of opium production, highland drug lords began to act as independent entrepreneurs responding creatively to market opportunities for their product. In Burma, for example, opium production increased exponentially from 550 tons in 1981 to 2,500 in 1989 through the efforts of leading drug lords like Khun Sa. Similarly, the emergence of Gulbuddin Hekmatyar as the dominant rebel leader in

Afghanistan created a parallel figure of power that could control much of the country's opium production, heroin processing and export (Coy, 2003).

The success of the narcotics business is heavily dependent on clever illegal trafficking operations, and its lucrative markets are in Europe and North America. According to a source, the narcotics business ranks as the world's most successful illegal enterprise, generating annual profits of approximately US$200 billion to US$300 billion. Earlier, most of these drugs used to travel through the Iran-Turkey route - that is, the so-called Balkan Corridor. However, after the sanctions imposed by Iran on heroin production, the most favoured trafficking route from Southwest Asia to the West has been through the Fergane Valley towards Russia and the Baltic States.

It is important to note that Pakistan's ISI supported drug smuggling as a means of laundering huge amounts of money to meet the heavy expenses of war on various fronts, including Afghanistan and India's states of Punjab and Kashmir. The Mujahideen of Afghanistan accelerated the production of heroin, which became a lucrative business for them. Besides the land route to Central Europe via the Balkans, the Indo-Pakistan border is also being used extensively for transporting narcotics, first to India and then to other countries. Militancy in the Jammu and Kashmir states of India and the smuggling of narcotics have mutually complemented each other's requirements - that is, the need for easy money.

In the context of the proxy war in Jammu and Kashmir, the ISI of Pakistan is using the narcotics trade to: a) generate funds to sustain militancy; b) weaken the strength of the populace in the border belt; c) win over the local youth to work as informers/suppliers in support of their cause; and, d) increase the level of criminal activities. According to reports, much of the heroin which originates in Afghanistan is transported to Peshawar via the tribal areas adjacent to the Afghan border. Most of the Afghan refugees are mainly involved in heroin trade as poppy growers, stockists, middlemen or international traffickers. In the border areas, armed guards of drug mafias act in the guise of rebels. They use the airport and sea route in Karachi to supply drugs to the northern countries. In fact, every month approximately 20,000 kg of hashish and cannabis leaves from here to different destinations.

Besides heroin, chemical precursors were also reported to be moving in from Central Asia. The chemicals were transported to the refineries in Pakistan and Afghanistan controlled by cartels based in Quetta, which in turn is tied up with Turkish and Iranian buyers, to route the heroin to the international market. These major drug traffickers are now associated with a number of factional groups in Afghanistan. Heroin labs were set up not only in Taliban-controlled areas but also in the Uzbek border area, which was controlled by Dostum. There were also organised drug cultivation areas controlled by Masood. Drugs from these areas left for Russia, China, Tajikistan and Kyrgyzstan.

Drug trafficking organisations also have strong linkages with the Gulf countries. Dubai seems to be emerging as an important centre for drug money laundering. Cannabis is usually transported from Afghanistan through Baluchistan to the Makran coast of Pakistan and from there, by ship, to the Gulf States and Europe. Cannabis is also increasingly transported through the Central Asian Republic by rail (Kartha, 1999).

In Pakistan, opium production increased throughout the 1980s until 1985, then declined in 1990, partly due to poor weather conditions and partly through crop eradication. However, the number of mobile labs producing heroin has reportedly increased to more than 200 in recent years. These labs were set for the purpose of converting opium to morphine and from morphine to heroin. These labs are mainly located in the border areas, particularly in the Khyber region adjacent to Afghanistan, and close to the highway connecting Khyber to the rest of the country. This is the chief production zone for the injectable variety of heroin, or heroin no 4 (Herald, 1990).

Interestingly, Pakistan's drugs bazaar is located just 7 miles from Peshawar, where US officials from the Drug Enforcement Agency (DEA) and Pakistan's Anti-Narcotics Board have their offices. In this bazaar, heroin is available in 50 kg sacks and hashish by the tonne. There are dozens of small markets just outside the tribal belt in NWFP, where the Pakistan Government's law is not effective. A journalist of the Herald journal has described the situation wonderfully. He writes:

> Pakistan's tribesmen, now shopkeepers, sit cross-legged, sipping tea and smoking hashish-filled Benson & Hedges cigarettes, as they display weapons that range from AK47s to anti-tank missiles. But hashish is only pocket money to these tribesmen who have built massive fortress-styled homes decorated with Italian marble. The real business is heroin (Herald, 1990).

The extent of drug smuggling carried out by the Pakistan militants in the Kashmir Valley can be judged from the fact that 19,450 kg of narcotics (heroin, charas, brown sugar, and so on) valued at Indian Rupees 20 crore, were recovered from Kashmir during 1997-1998 (Herald, 2003). Drugs were also seized from Pakistani nationals by the Border Security Force of India in the western fronts of Rajasthan and Gujarat. Pakistan is also providing shelter to drug lords and criminals who are working against India. Dawood Ibrahim, a criminal wanted by India, is operating from Pakistan, and supports terrorist groups such as Lashkar-e-Tayyiba (LET), against India. LET is a Sunni, anti-USA militant group formed in 1989.

According to a report, Dawood Ibrahim has been funding the increasing attacks in Gujarat by LET. This group, led by Hafiz Mohammad, was placed on the US State Department's list of officially designated terrorists groups in 2002. Washington's decision seems to be largely motivated by the arrests in the USA and Saudi Arabia of members of an Islamic paramilitary unit who received training in small arms machine guns and grenade launchers at a LET camp in northeast Pakistan. LET is said to be active in the Washington D.C. area and nearby Virginia state of USA. Documents released by the FBI in June 2003, state that some of those arrested even fought against Indian troops in Pakistan-occupied Kashmir. They were also funded by the Dawood Ibrahim syndicate to conduct terrorist activities in the Indian state of Gujarat. According to the US Treasury Department's fact sheet, the Ibrahim Syndicate is involved in the large-scale shipment of narcotics to the U.K. and Western Europe. Its smuggling routes from South Asia, the Middle East and Africa are shared with Al Qaida and its terrorist network.

The LTTE of Sri Lanka too has deeply penetrated the drug world, to augment its armed strength against the Sri Lankan Army. It has established a supply network for heroin from India to Western Europe. For the LTTE, trading in gold, laundering money and trafficking in narcotics brings in substantial revenues needed to procure sophisticated weaponry. The SAM missiles which the LTTE has procured from Cambodia, cost the LTTE US $1million each. The sophisticated arms used by the LTTE, including shoulder-fired surface missiles, rocket launchers and global positioning satellite systems are all purchased with drug money. What is more, LTTE suicide bombers have been trained in France and Britain, to fly light aircraft. These ultra lights do not carry sufficient metal for radar detection. Further, they could take off from a short runway.

India, which serves as a transit point for smuggling and contraband products, is also emerging as a big producer and consumer of drugs. The north Indian states of Madhya Pradesh, Rajasthan and Uttar Pradesh are heavy poppy growing areas, with a number of established heroin production units. New Delhi has been a major consumer and networking centre for the heroin products brought from these neighbouring states and transported to the other countries.

In the African continent, the role of East and Southern Africa in the supply and distribution of drugs has become more prominent in recent years. This was due to the general deterioration of law and order in this region. The growing role of criminal groups in Nigeria in international drug trafficking has also been a factor in this trend. In the Southern African region, South Africa is being targeted both as a major end user of narcotics as well as a transshipment point.

The Need for Regional Cooperation

From the above discussion, it is evident that the spread of illegal arms and drugs in the Indian Ocean Region has led to an alarming and volatile situation. It is difficult for a single country to handle and control this situation. As it is, the region has already seen the emergence of some successful regional organisations like the Association of South East Asian Nations (India is a dialogue partner of ASEAN), the Asia Pacific Economic Cooperation group, and several other regional organisations such as the BIMSTEC, and the Mekong Ganga Co-operation Project. What is noteworthy in this context is that all of these organisations have member states that are maritime countries, with the notable exception of landlocked Laos. In such a scenario, the Andaman and Nicobar Islands of India provide the most ideal logistical base from where sea power could extend its reach. It is all the more significant that all of these countries are critically dependent on the key 'Strategic Sea-Lane of Communication' (SLOC) of the Straits of Malacca. It may also be noted that unlike other regions, which have formed politico-security cooperation frameworks like the OSCE in Europe, the OAS in America, the OAU in Africa, the Indian Ocean Region lacks any overarching security framework at the moment. Until recently, a majority of the states in the region tended to rely on conventional means of ensuring security such as expanding national military strength and accretion to that strength through an alliance; in short, a regional variant of the balance of power approach.

However, the experiment with the Asean Regional Forum (ARF) shows that there has been some movement from a sole reliance on bilateral and multilateral military alliances to a more cooperative approach on security. Similarly, India, which has always believed in the principle of an independent security paradigm, has been a part of this regional cooperation on security matters. India has been conducting regular joint naval exercises and joint military training with Southeast Asian countries, both bilaterally and multilaterally.

It is also a manifestation of the Look East Policy that the scope of India's naval diplomacy has been extended further with the country conducting joint military exercises across the Indian Ocean with countries such as Vietnam and South Korea. India is also proposing that Japan and Vietnam along with it should be strategic partners in anti-piracy operations that will involve other Asian countries for joint military operations.

Presently, the United Nations Fund for Drug Abuse Control (UNFDAC), established in 1971, is the only international body entirely devoted to assisting governments in combating the production of drugs, drug trafficking and the use of illicit drugs. The Governments of South Africa and Mozambique, which are the most affected, are aggressively challenging the culture of violence, with strong support from their civil societies and the UN. The governments of both countries undertook joint operations to curtail the flow of apartheid-era weapons entering South Africa from former war zones in Mozambique.

The Organisation of African Unity (OAU) has taken other significant initiatives in 2000. It met in Bamako, Mali, in 2000, to develop an African Common Position on small arms and Light Weapons (SALW), in anticipation of the 2001 UN Conference. The OAU took cognisance of the earlier African initiatives, like the 1998 ECOWAS Moratorium and the 2000 Nairobi declaration, among other African regional initiatives, as a starting point. The Bamako Declaration put demand reduction strategies on the policy map. While carefully reaffirming the values of sovereignty, non-interference, and the right to individual and collective self-defence, it emphasised that "the problem of the illicit proliferation, circulation and drug-trafficking of small arms and light weapons sustain conflicts promotes a culture of violence has an adverse effect on the security and development and is both on supply and demand".

One effective way to tackle the increasing proliferation of small arms in the IOR is to encourage governments in the region to register every weapon produced in their respective countries. This would help them trace the source of the arms seized from terrorists and armed traffickers. Such arms seizures might reveal some politically embarrassing facts, but almost certainly this measure would help reduce the current proliferation of military hardware and its nexus with drugs in the Indian Ocean Region.

Fortunately, organisations in the region like the SADC and SAARC have initiated some programmes to counter these problems. For instance, in August 2001, the Southern African Development Community (SADC) adopted a protocol on the Control of Firearms, Ammunition and other Related Materials in the SADC

Region. Like the Bamako Declaration, the SADC Protocol focuses mostly on supply-side interventions, but also acknowledges the key relationship between limiting the availability of weapons and maintaining stable peace processes and post-conflict situations. Two priority areas for the SADC are cooperation with the United States in a joint working group on small arms, and continued cooperation with the European Union. In South Asia, SAARC has signed Conventions on Terrorism and Drug Trafficking and the Terrorists Offences Monitoring Desk and the Drug offences Monitoring Desk.

Thus, in the wake of all these initiatives in the global, national and regional levels on the proliferation of small arms and drug trafficking and the acrimonious discussion surrounding it, the timing is propitious for a rigorous re-examination of the scope of the problem and the political, strategic and economic mechanisms for dealing with it. Targeted empirical research could also make a useful contribution to the debate by examining in more detail the nature of the problems caused by this dangerous scourge.

References

Agha, Ayasha Siddiqua (1991), 'Setting an agenda for Regional De-Weaponisation', in Dipankar Banerjee (ed) *South Asia at Gunpoint: Small Arms and Light Arms Proliferation* Colombo, Sri Lanka: Regional centre For Strategic Studies 2.

G. Davidson (Tim) Smith, (1991) *Terrorism and the Rule Of Law: Dangerous Compromise in Colombia.* Commentary *No.13*, Strategic Analyst in the Analysis and Production Branch (RAP) of CSIS, Canada

Chalk, Peter, (1996), *'Liberation Tigers of Tamil Eelam: International Organizations and Operation, Commentary No. 77.* Canadian Security Integence Service, March 17, 2000.

Coy, Alfred M. (2003), *The Politics of Heroin: CIA Complicity in the Global Drug Trade,* New York, Lawrence Hill Books, Revised.

Fung, I. R. (2000), *'Small Arms and Light Weapons in Africa'* Illicit Proliferation, Circulation and Trafficking', in *Proceeding of the OAU Meeting and International Consultants,* Pretoria. compiled by Eunice Reyneke, Pretoria Intitute for Security Studies

Gallaghar, James, J. (1992), *Low Intensity Conflict: A Guide for Tactics, Techniques and Procedures*, New Delhi: Lancer Publishers.

Herald, (Pakistan), April, 1990.

Herald, (Pakistan), November, 2003.

Kartha, Tara (1999), *Tools of Terror: Light Weapons and India's Security*, New Delhi: IDSA.

Klare, Michael (1995), 'Light Weapons Diffusion and Global Violence in the Post-Cold War era', Basic, *IDSA.*

See *Special Focus on Illegal Narcotics, South African Journal of International Affairs,* Vol 5 (1). Summer 1997.

Rao, P. V. (2003), 'India and Regional Cooperation: Multiple strategies in an Elusive Region', in P. V. Rao (ed) *India and Indian Ocean*, New Delhi: South Asian Publishers.

Reyneke Eunice (2000), *Small Arms and Light Weapons in Africa:* Proceedings of the OAU Experts Meeting and International Consultation, Institute for Security Studies.

Rubin, Barry, ed. (1991), *Terrorism and Politics*, London: Macmillan.

Singh, Jasgit, ed. (1995), *Light Weapons and International Security*, New Delhi: Pughwash, BASIC, IDSA.

Smith, Chris (1993), *The Diffusion of Small Arms in Pakistan and Northern India*, London: Brassey.

Subir, Bhowmik (1996), *Insurgent Cross Fire: Northeast India*, New Delhi: Lancer Publishers.

The Hindu, 18th August 2000

United Nations Document (1997), A/52/298, 27th August.

United Nations Document (2004).

Vaher Ado, United Nations Document (2000). GA/DIS/3177

Chapter 11

Container Security Initiatives: A South Asian Perspective

Vijay Sakhuja

Introduction

Containerised cargo systems have today emerged as the most convenient and cost effective mode of transporting large volumes of goods. Nearly 90 per cent of the world's cargo movement takes place through containers and over 200 million cargo containers move between major seaports each year. According to the latest forecast released by the container data agency, BRS-Alphaliner, the number of container ships worldwide is set to grow by about 13.7 per cent between now and 2007. By the end of 2007, the global fleet of containerships is expected to be 4271, up from the current 3362 vessels totaling 7.29 million TEUs.[1]

Meanwhile, according to the annual UNCTAD 2004 report, average export and import growth rates for 40 selected Asian economies reached 14.8 per cent in 2003, and Asian countries were major players in world maritime transport, having sizable shares of several activities. These countries accounted for 35.8 per cent of container ship ownership, 45.7 per cent of container ship operation, 60.4 per cent of seamen, 62.3 per cent of container port throughput, 64.7 per cent of container port operators, 83.2 per cent of container ship building and 99 per cent of ship demolition.[2]

While this is a welcome development, container shipping appears to be quite vulnerable and has the potential to be the 'Achilles Heel' of seaborne trade. Shipping containers can be used as means of transporting terrorists as well as weapons of mass destruction and can also serve the purpose of facilitating illegal trade. When used for such purposes, container shipping has the capacity to disrupt and even destroy maritime enterprise and threaten the peaceful use of the seas.

This chapter begins by tracing the history of container trade. It stresses the vulnerability of container trade to terrorism and highlights the technological solutions required for the safety and security of containers. It also examines the competing requirements of maritime trade and security. Finally, the chapter discusses South Asian initiatives with regard to the Container Security Initiative.

Malcolm MacLean and the Origins of the Container Trade

A trucker by profession, Malcolm MacLean can be credited for being a pioneer in introducing container transportation. In 1931, the first container truck went into operation on the roads in the United States. Over the years, MacLean increased the number of trucks into a major fleet of several hundreds. He began strategising ways to overcome transportation bottlenecks and argued that if the trailer could be moved directly to the carrier, transportation could be efficient both in terms of cost, time and safety. This was particularly so at the point where the cargo/consignment is transferred from one mode of transportation to another D that is rail to road, from truck to ship and vice versa. By 1956, he bought a small tanker company named Pan Atlantic and modified two of its ships to carry 58 detachable trailers. In order to disengage the trailers, he removed the wheels and strengthened the sides. The first of these converted ships sailed from New York harbour to Houston, and containerisation became a reality. However, it was only in 1966 that the concept was utilised in international trade on the Japan-California-Europe route. MOL's first container ship, the 797-TEU America Maru, was launched in 1968.

Today, we have container vessels like the Regina Maersk capable of transporting as many as 6600 TEU containers. From vessels that used to carry 226 TEU's in 1957 there are today vessels that can carry 6600 TEUs. Maersk Sealand alone has around 21 vessels that can carry over 6000 TEUs. Their 'S'-Class Post Panamax vessels can carry 6600 TEUs. Other lines having over 6000 TEU vessels in their fleet are MSC, P&ONL Hanjin, Hyundai Merchant Marine, and CMA-CGM. The world fleet at present consists of 32 vessels of 6000 TEUs and above, with another 40 on the order books and many more to follow.

Meanwhile, reports suggest that four of the world's largest declared capacity container ships capable of carrying 10,000 TEUs are to be built in Korea at Hyundai Heavy Industries for the China Ocean Shipping Corporation (COSCO).[4] The vessels will be delivered between late-2007 and mid-2008. The 10,000 TEU container ships ordered by Cosco are a step towards the 12,500 TEU limit.

Competing Requirements: Trade and Security

During the last decade, the international community has been discussing the impact of globalisation on international relations, economy and trade. It is generally accepted that the world as a whole has benefited from expanded markets and free trade that emerged due to the rapidly growing global economy. The economic interaction, partly driven by technology and the cornucopia of ideas, is perhaps shaping today's global economy. There is global interconnectivity and we are witnessing large volumes of goods and services moving rapidly across societies. By and large, globalisation has helped humankind to attain a degree of economic prosperity. While there are proponents of globalisation there are also detractors. It appears that only a small minority has rejected globalisation and even fewer warn of its negative impacts. While a globalised world economy calls for open markets, more international trade/commerce and greater openness, the security of the globalised economy also needs to take priority. The question is how do we manage the competing requirements of security and economic vitality in this new world? The answer lies in prudence and innovation. The requirements of security therefore have to be carefully chosen so as not to impede economic growth resulting in a potential slowing down in the flow of commerce, but, at the same time, requirements have to be stringent and comprehensive enough to secure and defend trade against a new array of threats in the post-9/11 world, ranging from nuclear/biochemical attacks to suicide boat attacks like the *USS Cole* and *MV Limburg* incidents.

Security Threats From Container Trade

Although the introduction of shipping containers in 1956 was a revolution in the process of shipping vast amounts of goods across the ocean, these boxes, which can be easily shifted to railroad cars or trucks, have emerged as the soft underbelly of sea trade and a potent tool for terrorists. It is now widely accepted that maritime infrastructure is the soft underbelly of states that can be attacked with little effort. It is also believed that shipping containers are a safe sanctuary for terrorist activities. Large container vessels dock at ports almost every day and the cargo is rarely subjected to thorough inspection.

As a matter of fact, shipping container trade has now come to the frontline in the war on terrorism. Clearly, one of the biggest challenges is to know what is inside the container. Fears of containers being used as a "dirty bomb" or as a "slow moving cruise missile"[5] have heightened. Besides, containers are being used as cruise cabins by terrorists, illegal immigrants and as a means to carry out illegal activities at sea. These steel boxes have exposed the vulnerability of maritime shipping. After the container door is closed and sealed (under varying degrees and seriousness of inspection), these boxes move into seaport terminals aboard container ships, trains, trucks and even aircraft, with little or no information about their true contents.

Some time ago, the freighter, *Palermo Senator,* was ordered to remain six miles off the coast of New Jersey. The vessel was scheduled to offload 655 containers in Port Elizabeth. The US Coast Guard agents detected traces of low radiation[6] while

(b) Pre-screening those containers that pose a risk at the port of departure before they arrive at US ports.

(c) Using detection technology to quickly pre-screen containers that pose a risk.

(d) Using smarter, tamper-evident containers. [13]

Twenty countries have committed to participation in CSI. A total of 36 CSI ports are currently operational. They include:

Halifax, Montreal, and Vancouver, Canada (03/02); Rotterdam, The Netherlands (09/02/02); Le Havre, France (12/02/02); Marseille, France (01/07/05); Bremerhaven, Germany (02/02/03); Hamburg, Germany (02/09/03); Antwerp, Belgium (02/23/03); Zeebrugge, Belgium (10/29/04); Singapore (03/10/03); Yokohama, Japan (03/24/03); Tokyo, Japan (05/21/04); Hong Kong (05/05/03); Gothenburg, Sweden (05/23/03); Felixstowe, United Kingdom (U.K.) (05/24/03); Liverpool, Thamesport, Tilbury, and Southampton, U.K. (11/01/04); Genoa, Italy (06/16/03); La Spezia, Italy (06/23/03); Livorno, Italy (12/30/04); Naples, Italy (09/30/04); Gioia Tauro, Italy (10/31/04); Pusan, Korea (08/04/03); Durban, South Africa (12/01/03); Port Klang, Malaysia (03/08/04); Tanjung Pelepas, Malaysia (8/16/04); Piraeus, Greece (07/27/04), Algeciras, Spain (07/30/04), Nagoya and Kobe, Japan (08/06/04), and Laem Chabang, Thailand (8/13/04). [14]

As far as China is concerned, the port of Shanghai became the 36th operational port to target and pre-screen cargo containers destined for US ports. [15] According to the agreement, the US Customs and Border Protection (CBP) will deploy a team of officers to be stationed at the port of Shanghai to target maritime containers destined for the United States. Shanghai Customs officials, working with CBP officers, will be responsible for screening any containers identified as a potential terrorist risk. CBP will also share its information and pre-arrival data on a bilateral basis with its CSI partners. Sharing of information is intended to be a reciprocal process. [16]

Technological Solutions

One of the biggest challenges facing security agencies is to know what is inside each container. After the container door is closed (under varying degrees and seriousness of inspection) it changes hands several times at factories/warehouses, seaports/airports, aboard container ships, trains, lorries, before it reaches its destination. Examining all containers violates an age-old axiom in the security field that if "you have to look at everything, you will see nothing". There is thus a need to install equipment that provides a comprehensive and a credible means of determining the contents of a container.

The solution lies in developing such technologies that are non-intrusive, safe for operators, cost effective, highly reliable and, more importantly, speedy in order to avoid unwarranted interruptions to the flow of commerce. Non-intrusive technologies were developed during the 1990s to inspect cargoes both at seaports and land border crossings. These use a variety of radiations that penetrate through the container to produce an image of the contents of a sealed shipping container or any vehicle. The scanning radiations currently in use are: x-ray, gamma ray, and high-energy neutron. The non-intrusive scanning systems are based on X-rays, Gamma Rays and Neutron Analysis technologies.[17] These are used for the inspection of vehicles and containers for the detection of contraband and cargo verification. Mobile x-ray systems, due to radiation safety limitations, work on restricted level of x-ray energy which can penetrate cargo equivalent to a density of approximately 220 mm steel plate and resolution up to 2.5 mm steel/copper at the centre of the cargo. For the other configuration, wherein major radiation shielding is provided to the x-ray source, the penetration can be the equivalent of 300 mm of steel plate with a resolution of 3.0 mm. However, x-ray systems are very heavy and have limited mobility. For achieving higher penetration levels of up to 400 mm of steel plate that will enable scanning of dense cargo containers, high energy levels of 9.0 MeV are used. In this case, due to radiation safety limitations, the system is enclosed in radiation proof building.

The mobile Gamma Ray systems are lightweight and use radioactive Cesium or Cobalt as a source, which generates a narrow vertical fan beam of gamma rays directed at highly sensitive detectors. The Cobalt 60 Source is preferred to Cesium because it is safe and has a higher efficiency. These systems are relatively much lighter compared to the mobile x-ray systems, but have a somewhat lower penetration of around 170 mm of steel with good contrast sensitivity. As such, these systems are ideal for empty and up to medium density cargo containers and the comparative cost is substantially lower than x-ray systems.

The gamma ray scanning system is also available in the configuration of a static portal-based system through which the container is made to pass. The mobile truck has a compact radiological threat identification system (RTIS) to allow positive isotopic identification of Radiological Dispersal Devices (RDDs) and neutrons from fissile material that may be concealed in containers. Thus, the gamma-ray radiography system can also identify weapons of mass destruction.

Both the x-ray and gamma ray scanning systems work well with cargo that has a definite shape and is recognizable. Their efficacy has been suspect, however, in the case of drugs, explosives and other materials that can be packed in a variety of containers of different shapes. For instance, plastic explosives can be packed in any type or shape of container and transported without being detected. Similarly, home made ammonium nitrate and fuel oil bombs that terrorists use can be transported in simple plastic drums. It is extremely difficult to differentiate a full barrel of explosives from one with legitimate fruit puree or industrial chemicals. Narcotics can also be packed in variety of containers, toys, food packaging materials and in refrigerated cargo like fish and meat. It is very difficult for x-ray systems to detect such contraband.

In the 1990s, neutron-based cargo inspection devices emerged in the US. The US Government worked in partnership with private industry to develop scanning systems to overcome the drawbacks in the x-ray imaging systems. The remedy was found in neutron inspection systems that use neutrons to stimulate material specific signals from the inspected object. These signals are determined by the cargo's elemental composition. These data are captured and passed through a computer library data base of materials. If there is a match, the system sends warning signals to the operator. The detection is automatic and does not rely on the human operator. Pulsed Fast Neutron Analysis (PFNA) is perhaps the most advanced neutron scanning technology that is capable of detecting contraband concealed in fully-loaded cargo containers and trucks. The system offers a three dimensional picture of the cargo which is compared with the computer's database of known material threats. The PFNA throughput is less than x-ray scanning through put. To obtain optimum performance, the x-ray scanning system and the PFNA system are best used in tandem. The x-ray system carries out a relatively fast scanning and if a particular area of a container is suspect, the container is passed through a PFNA scanning, which is focused on to the suspect area which improves the throughput considerably.

The South Asian Context

A country's response must be commensurate with its level of international trade. According to a recent report, worldwide container traffic was projected to grow from 87.1 million TEUs in 2003 to 94.4 million in 2004 and to 104 million in 2005. By 2010, it is expected to hit 400-460 million TEUs and 510-610 million in 2015. Asia will remains the dominant player in the global shipping industry, with its container trade projected to grow at an annual average rate of seven percent between 2000 and 2007.

India
Asia's shipping trade is expected to grow sharply, led by China and India. India is yet to take any initiative with regard to container security. Not so long ago, bomb disposal squads of the Indian Army and the National Security Guards diffused live rocket shells from a scrap consignment that arrived at the Bhushan Steel Company

from Iran. This US$25,000 consignment, loaded in shipping containers, left Bander Abbas in Iran on board *M V Kuo Long*. The containers were bound for the Inland Container Depot (ICD), Tughlakabad, near Delhi via Mundra port in Gujarat.

Close on its heels, in May 2005, the Mumbai police seized a Rs1-crore haul of small arms. These had been cleverly packed in one of the 20 barrels of industrial grease carried in a container and imported by a biotech and poultry food company. The consignment originated in Bangkok and included 34 Webley Colt and Wesson revolvers, three powerful pistols, a silencer and some 1,283 rounds of [18] ammunition. Interestingly, eight containers, each with several barrels, were addressed to the agro-biotech firm. Four of the containers were delivered and the fifth was inspected by the Customs and Directorate of Revenue Intelligence and was scanned using the hi-tech x-ray facility. The other three containers are reportedly missing.

In December 2004, India and the United States signed a bilateral Customs Mutual Assistance Agreement (CMAA). The CMAA was conceptualised in 2001 and approved by the Indian Cabinet in 2003. It establishes a formal mechanism for the exchange and sharing of intelligence and information between the two Customs Administrations - that is, the Central Board of Excise and Customs, in the Department of Revenue of the Indian Ministry of Finance and Immigration and the Customs Enforcement (ICE), and Customs and Border Protection (CBP), under the US Department of Homeland Security. This bilateral agreement goes beyond the already existing working relationship between the custom authorities and aims at prevention, investigation and repression of customs offences. Among other issues, the agreement envisages that the two custom authorities share intelligence and information on terrorist activity, trade fraud, narcotics trafficking, smuggling of weapons of mass destruction and container security violations.

The biggest fear among the inspection agencies is that, with the increasing globalisation and trade, container frauds will be on the increase. Revenue agencies are losing huge sums of income through container fraud. Reportedly, in 2000, the global loss in revenue from tariffs and excise taxes resulting from container trade fraud alone was estimated at about $US170 billion. Container fraud essentially relates to a deliberate act on the part of consignor/consignee to misrepresent cargo to avoid duty or specific prohibitions on goods. The latter relates to smuggling, a practice that has been in existence in the world of trade from time immemorial. In addition, importers and exporters evade customs duty by under-valuation or false commodity classification. Some of the higher-value finds have been machine parts declared as clothing and finished products as complete knockdown (CKD) kits. There is no denying the fact that tariff collections are a significant source of government revenue. By some estimates, developing countries derive between 40% and 80% of their revenue through duties.

According to the Defence Minister, Mr Pranab Mukherjee, India is planning to buy three to five gamma ray-based cargo container scanners. One such system is in operation at Nhava Sheva Port at Navi Mumbai and has so far performed to satisfaction. In the Indian context, the government is committed to enhancing international trade and has even announced plans to invest US$10billion from its foreign exchange reserves for the development of its container terminal facilities at ports across the country. This is part of the new national maritime policy that envisages raising the capacity of ports and promotes transshipment cargo with an emphasis on public-private partnerships. Private sector participation is aimed at infrastructure and commercial activities within ports and public investment is directed at common user facilities at ports.

Bangladesh
A three-member expert team from the US visited the Chittagong port in May 2003 for a feasibility study and, in a sitting with the CPA chairman, came to the conclusion to set up a CSI in the port. [19]

In September 2003, while making a presentation on a Customs Trade Partnership Against Terrorism (CTPAT) at Dhaka Chamber, the Assistant Customs Attache of the US Embassy in Singapore, Peter R Darvas, had noted that the US Bureau of Customs and Border Protection (CBP) was not pressing for the setting up of a security scanning machine at Chittagong Port to check export consignments destined for the United States.[20] He noted: "We don't have any plan so far to have a container security initiative (CSI) in Chittagong Port. We aren't setting any scanner here."

However, by June 2004, according to *The Daily Star*, Bangladesh had plans to deploy container scanning machines at the port of Chittagong and expects to put at least one into operation by the end of 2004. This announcement was made after Bangladesh signed an agreement with the US to "extend full cooperation to the US in identifying individuals or groups suspected of working against US interests".[21]

Sri Lanka
In June 2003, Sri Lanka and the United States signed a Declaration of Principles (DOP) for the Container Security Initiative. Under the DOP, Sri Lanka Customs and the US Customs and Border Protection are to exchange information and work closely together to identify and screen high risk containers bound for the US. A small number of CBP officers will be deployed at the Colombo port to work jointly with Sri Lankan counterparts to pre-screen and target high risk cargo containers. Sri Lanka is strategically located at a key crossroads in the global trading system with a high potential for detecting items of concern. Approximately 70 percent of the containers handled in Sri Lanka are transshipments.

Pakistan

In February 2005, the Pakistan government announced that it intends to introduce the requirements of the Container Security Initiative (CSI) to all exporting containers. It was also announced that an inter-ministerial committee comprising of representatives from the ministry of commerce, CBR and the ministry of interior headed by the federal minister of ports and shipping, had been constituted for the implementation of CSI. The federal secretary, ports and shipping, warned that: "The world has changed dramatically since the tragic events of 9/11; we now live in an era of suspicion and fear of terrorist action on land, sea and in the air". Concurrently, the Pakistani authorities are engaged in consultation with stakeholders for implementation, but the pace of formulation of CSI is very slow. It is also believed that, since 2001, besides regular federal American security agencies, a large number of *secret* agents appointed by the American security agencies are monitoring container movements from Pakistan and the collected information is sent to the American authorities on a daily basis.

Pakistan's National Logistic Cell (NLC) has plans to install scanners at ports to facilitate the detection of potential problems. It is believed that the initiative will ensure the country's image overseas besides facilitating speedy clearance. The NLC is expected to handle the scanning operations on a professional basis rather than commercially.

Meanwhile, scanners of Chinese origin installed at Port Qasim, have not performed to the requisite manufacturing specifications. There is a fear that the Chinese scanners will not be acceptable to American security authorities. The installation of another set of scanners at Karachi Port Trust (KPT) has been delayed due to ongoing wrangling between senior members of the KPT and the National Logistic Cell.

Concluding Remarks

The threat of terrorism has placed special emphasis on trade and transportation security. Containers have emerged as the classical Trojan Horses. Governments and industries have to find new ways to respond to the threat of containers being used as tools for terrorism, illegal activities, drugs and arms smuggling and even for transporting weapons of mass destruction. Container security is technology-intensive. There is thus a need to install equipment to scan containers to examine hidden compartments and inspect the contents without unloading the container. However, to enable a safe system of commerce, a comprehensive and credible approach to security is essential.

It is true that the South Asian container trade is growing both in volume and value but only one South Asian seaport will soon be CSI-compliant. Regional governments have acknowledged the need for the deployment of scanning equipment and are conscious of the fact that the consequences of failure to detect contraband can have devastating security and revenue repercussions. This failure also has the potential to slow down the flow of commerce. The remedy lies in stringent and comprehensive detection measures to secure and defend maritime trade.

Endnotes

1) For more details see < http://www.alphaliner.com>

2) 'Review of Maritime Transport 2004' available at <http://www.unctad.org/ Templates/WebFlyer.asp?intItemID=3368&lang=1>

3) See 'Containerisation' at <http://www.choicegroup.co.in/html/cntrzation.htm>.

4) See Lloyd's Register to class world's first 10,000 teu container ships, February 21, 2005 ,at <http://www.lr.org/news/press_releases/2005/pr_0221_cosco.htm>

5) According to Irvin Lim , in an age of unrestrained terrorism shipping containers can be likened to Trojan [Sea] Horses or slow moving cruise missiles.

6) See 'Stowaway Terrorists Steal into America by Sea Container' at DEBKA-Net-Weekly Intelligence Report, 18 June, 2002 at http://www.debka.com/LADEN/ body_laden.html.

7) See 'Terrorist In A Box : Business-class Suspect Caught In Container' at <http:// hypocrisytoday.com/stowaway.html>.

8) See 'Port of Entry Now Means Point of Anxiety' at <http://college4.nytimes.com/ guests/articles/2001/12/23/892576.xml>

9) See ' Stowaway Qaeda Terrorists Steal into America by Sea Container' DEBKA-Net-Weekly Intelligence Report, available at < http://www.Halsteel.com>

10) See 'Threat Assessment: ABC News Smuggles Uranium Into United States' at < http://www.Salon.com dated September 6, 2002. According to the story, out of the 1,139 containers on the vessel used to ship the depleted uranium to New York, the ABC package was one of less than a dozen identified for further inspection before the ship entered port, said U.S. Customs Service spokesman Dean Boyd. Inspectors used X-ray equipment and a radiation detector to check the ABC package, which was found to not pose a threat, he said. The suitcase of depleted uranium, however, would give off about the same amount of radiation as a package of enriched uranium would if shipped in a lead-lined case, Ross said. The container should have been opened and examined, he said. "They missed it," Ross said. "They could say that it was no danger, which is true because we made sure there was no danger. But I think that misses the point" U.S. officials are angry over the time spent on ABC News's operation, AP reported.

11) See 'Container Cargo Theft Highlights Importance of Verifying Contents in All Port or Ship Security Strategies' dated 20 December 2002 at < http://www.iccwbo.org/index_ccs.asp>

12) The CSI is now under the U.S. Customs and Border Protection (CBP), Department of Homeland Security.

13) US Customs and Border Protection, US Department of Homeland Security, CSI Fact Sheet, January 7, 2005.

14) Ibid.

15) China Implements Container Security Initiative at Port of Shanghai to Target and Pre-Screen Cargo Destined For U.S. at < http://www.customs.gov/xp/cgov/newsroom/press_releases/0042005/04282005.xml>

16) Ibid.

17) Details of the X-rays, Gamma Rays and Neutron Analysis technologie scanners have been obtained from the ECIL-Rapiscan database.

18) http://www.outlookindia.com

19) "Delay in Setting up CSI Resented", available at www.independent-bangladesh.com/ news/nov/10/10112004ct.htm.

20) "US Not to Install Scanner at Ctg Port" available at<www.newagebd.com/oct1st03/011003/busi.html>

21) "Bangladesh to Deploy Container Scanners", available at <strtrade.com/wtf/home.asp?pub=0& story=16854&date=6%2F3

Chapter 12

Environmental Security Issues in the Indian Ocean: A Preliminary Analysis

Mohd Nizam Basiron and Mokhzani Zubir

Introduction

Environmental security in the Indian Ocean is an interplay of its ecological functions, internal humanistic influences and external pressures for more intensive exploitation of fisheries and other resources. The dependence of the coastal communities on the Indian Ocean for sustenance and the relative economic underdevelopment of a large part of the Indian Ocean Region means that protecting and conserving the environment and resources of the Indian Ocean would ensure human security, particularly food security. However, a rise in the surrounding population and an increase in the presence of foreign fishing fleets pose a growing threat to the fisheries resources of the Indian Ocean. In addition, all three important ecosystems in the Indian Ocean region - coral reefs, seagrass beds and mangroves - are threatened by a combination of natural bleaching, increasing population and over-exploitation of resources respectively. Other environmental issues include loss of biodiversity, especially endangered species such as turtles and whale sharks and pollution from oil exploration activities and maritime transportation.

Environmental Security in the Context of the Indian Ocean Region

Environmental security is a broad term often used to denote a complex relationship between the environment and all levels of human society from communities to nation-states. This relationship concerns the impact of humankind on the environment and vice-versa and includes issues such as the detrimental impact of human activities on the environment and on the reverse side of the coin the impact of environmental changes on human beings . This relationship can be observed throughout the world but is more apparent in the Indian Ocean Region because of the population in the area which has doubled to two billion from 1950 to 1998; its large fishery which includes fifteen million active fishermen (out of the world total of thirty five million) ; and the relative economic underdevelopment of a large part of the region which means that there is strong dependence on agriculture and fisheries for sustenance.

This situation means that any decline in the quality of the Indian Ocean environments, be it in terms of water quality, fisheries resources or marine ecosystems and habitats, could adversely affect the sustenance and livelihood of coastal communities. There is evidence to suggest that this is happening in theIndian Ocean. This chapter explores some of the issues which affect environmental security in the Indian Ocean Region and their implications for environmental management in the area.

Prospects for the Indian Ocean Environment

Because of its sheer size and the varying levels of development among its littoral states, it is difficult to say with certainty what the overall outlook for the Indian Ocean environment is. However, two documents published by the United Nations Environment Programme (UNEP) in 1989 and 2004 outline key issues that face countries in the East African region of the Indian Ocean and the island states in the Indian Ocean respectively. In a nutshell, these reports indicate key marine environment issues as being loss of marine ecosystems and habitats; decline in endangered species; over-exploitation of fisheries resources; and pollution from land-based and marine-based sources [3]. This section of the chapter provides an elaboration of the broad issues identified in the two reports that have the closest links to human security.

Depletion of Fisheries Resources in the Indian Ocean

The Indian Ocean has the second largest fishing fleet in the world after the Pacific Ocean [4]. It is also seeing an increase in the presence of fishing vessels from extra-regional states, including Japan, the Russian Federation, South Korea and Taiwan, which focus on the capture of tuna and shrimp. The presence of high numbers of local and foreign fishing vessels have contributed to the steady increase in the total fish catch in the western and eastern Indian Ocean Region which has increased from 1.2 million tonnes to 3.9 million tonnes and from 1 million tonnes to 4.7 million tones respectively between 1970 and 2000. This has created tremendous pressure on the fisheries resource with the Food and Agricultural Organisation (FAO) expressing concern over the status of tuna resources and fisheries resources in the deep waters of the Indian Ocean [5]. Given the excess capacity of fishing fleets in countries like India, which accounts for most of the eastern Indian Ocean landings, and the small percentage of aquaculture production in the region, it is unlikely that the pressure on the Indian Ocean fisheries will be alleviated in the near future.

The pressure of over-fishing is also felt by non-commercial and endangered marine species. One disturbing feature of the Indian Ocean fishery is the high level of discards amounting to approximately 4.3 million tonnes in 2002. These discards include endangered marine species such as turtles. A recent phenomenon, which has also raised concern, is the harvesting of whale sharks off the coast of India. It has been reported that, in 2001, a total of one thousand whale sharks were caught for their oil and fins, prompting the formulation of laws to ban such practices [6].

Within the Indian Ocean Region, the southwest sub-region appears to be most adversely adversely affected by over-fishing, with seventy-five percent of fisheries resources being exploited at the maximum biological productivity level, mostly by external fishing vessels from Spain, Taiwan, Japan and Uruguay [7]. This has prompted the establishment of the South West Indian Ocean Fisheries Commission (SWIOFC) with the objective of assisting countries to manage the resources in a sustainable manner and to complement the work of existing bodies such as the Indian Ocean Tuna Commission (ICTC). The establishment of such a body, however, needs to be complemented by an overall improvement in the capacity of Indian Ocean states to manage their fisheries resources to ensure sustainability and safeguard the interests of coastal fisheries communities who depend on the Indian Ocean for sustenance and livelihood.

Developing states are not the only victims of the vagaries of industrialised fishing states as shown by the 2003 incident involving the Uruguayan trawler *Viarsa* [8]. The poaching of the Patagonian toothfish by the Viarsa and the subsequent six thousand kilometer pursuit may not have happened in the Indian Ocean but is ample illustration of the difficulties and costs involved in enforcing fisheries regulations over a wide expanse of sea area such as the Indian Ocean. Given this situation, a regional body such as the SWIOFC could contribute significantly towards regulating the non-Indian Ocean fishing vessels which are operating in the region to ensure that benefits from the exploitation of the resource are also extended to Indian Ocean countries.

Because the state of fisheries resources in the Indian Ocean is so closely linked to the well-being of its coastal population, its depletion could lead to a situation of resource scarcity. In an open access fishery such as the Indian Ocean this could result in conflicts between artisan fishermen who fish for sustenance and industrial fishing groups who fish for economic and financial purposes. A decline or lost of food security could result in the need for foreign food aid. For example, in 1998, both Madagascar and the Comoros received substantial food aid amounting to 26,000 tonnes and 26,000 tonnes respectively [9].

Marine Pollution

The Indian Ocean Region experiences both land-based and vessel-based pollution. Vessel-based pollution results mainly from the heavy tanker traffic which traverses the Indian Ocean which sits astride the two main east-west sea routes [10]. While the exact number of vessels passing through the Indian Ocean is not known, in 1989, an average of 225 tankers passed through the waters of East Africa daily consisting of both medium-sized tankers and very large crude carriers [11]. With increasing demand for energy it would be safe to assume that the number has also increased, thereby increasing the risk of oil spills in the area. Fortunately for the Indian Ocean, only two major oil spills have been recorded in the area - the World Glory in 1968 and the Katina P in 1992, each involving 14.2 million gallons and 15 million gallons of oil respectively [12]. Notwithstanding this, like the Strait of Malacca, the spectre of a major oil spill continues to demand the attention of the governments in the Indian Ocean Region, which unlike the littoral states of the Straits of Malacca, may not have the resources to address major oil spills.

Land-based pollution presents a perennial problem for countries worldwide. Unlike vessel-based pollution, land-based sources of pollution are usually more dispersed and harder to pinpoint. Among the island states of the Indian Ocean, land-based marine pollution originates mainly from agricultural, municipal and domestic sources. While industrial activities remain undeveloped, they are growing and will constitute future sources of pollution [13]. The situation is equally depressing among the larger Indian Ocean countries. In India, for example, only cities with populations larger than one hundred thousand have access to primary sewage treatment, resulting in the discharge of over 18 million litres of untreated or partially treated sewage into the environment in 1994 . The problem is widespread throughout this region which has the highest density of coastal population in the world at 135 people per square kilometre.

Destruction of Ecosystems and Habitats

The Indian Ocean hosts at least three important coastal and marine ecosystems - namely mangroves, seagrass beds and coral reefs. The Bay of Bengal and the Andaman Sea have among the highest diversity of mangroves in the world, while seagrass diversity is high in parts of the eastern African coast and the Andaman Sea. These ecosystems support the region's fisheries by functioning as nurseries and feeding grounds for many commercial species and sustain the livelihood of coastal communities along the Indian Ocean [15].

The marine ecosystems of the Indian Ocean, however, are being threatened by a number of factors. Primary among these are human activities and in the east African region, sewage discharge and over-exploitation have been added to destructive agricultural practices and deforestation as the primary causes of coral reef degradation [16]. The situation in Sri Lanka is not much different and the long list of causes of coral reef degradation includes destructive fishing practices, over-exploitation, sedimentation and coral mining [17]. Of the 36,100 square kilometres of coral reefs in the Indian Ocean, twenty five percent have been categorised as being under the high risk category, while another twenty nine percent is under the medium risk category. The World Atlas of Biodiversity describes the Indian Ocean as an "endemic rich area at higher risk" [18] meaning that there is a high diversity of species not found elsewhere, but that there is an equal level of risk to the biodiversity. Not all reef destruction in the Indian Ocean has resulted from human activities. The recent El-Nino phenomenon caused coral bleaching in many parts of the Indian Ocean with the Comoros reporting bleaching to as high as fifty five percent of its coral reefs [19].

The destruction or modification of marine and coastal ecosystems in the Indian Ocean could have adverse effects on the region's fisheries. Small coastal communities that depend on coastal and marine ecosystems for food would be severely affected by degradation or loss of coral reefs and mangroves, for example. To partly address the problem, governments in the Indian Ocean Region have established 66 marine protected areas covering an area of 15,100 square kilometres [20]. The setting up of these areas will contribute significantly towards the protection and conservation of marine ecosystems and ensure that the region's important fishery is sustained. One such protected area is the Indian Ocean Sanctuary which was established in 1979 and covers the entire stretch of the region and continues to protect Blue, Humpback and Sperm whales from whaling activities [21].

Natural Disasters

No discussion on environmental security in Indian Ocean would be complete without some mention of the December 26 tsunami. The devastation caused by the incident has been widely reported and publicised. What is equally important is how ill-prepared the region was in predicting the incident and minimising its effects on its coastal population. However, natural disasters are not uncommon to the region which has had its share of floods, typhoons and droughts. What is significant is the magnitude of the damage caused and its impact on the Indian Ocean community. The worst affected are the coastal communities, particularly small-scale farmers and fishing communities and small island states such as the Maldives which estimated that two-thirds of its population was affected by the tsunami [22]. Fishing communities throughout the Indian Ocean Region suffered tremendous material damage. In Sri Lanka, close to 20,000 boats were destroyed or damaged with repairs and replacement costs estimated at 75 million dollars [23].

The tsunami incident also highlighted the importance of protecting coastal and marine ecosystems, particularly mangroves and coral reefs which also functions as coastal protection from waves. This has prompted the government in Malaysia, for example, to issue directives to replant cleared mangrove areas and protect existing areas. Equally important is the need for an early warning system, and, after much discussion, the government of Indian Ocean countries have finally agreed on a tsunami early warning system for the region. The system will be based on a network of national warning system connected to sensors at sea [24].

Conclusions

It is difficult to make a general statement about the state of environmental security in the Indian Ocean because of the diversity of its littoral states in terms of population size, economic development and impact on the environment. There are, however, particular areas which appear to be coming under tremendous pressure from fisheries activities such as the western Indian Ocean Region. In terms of pollution, most of the Indian Ocean, apart from Australia, is beset by serious land-based pollution problems because of the overall lack of sanitary facilities and the high population density in the coastal regions.

More importantly, it should be noted that environmental security in the Indian Ocean Region is inextricably linked to human security. A number of factors point towards this interdependence. Firstly, while the consumption of fisheries resources by the population is low compared to Southeast Asia, the pressure on the resource is significant given the sheer size of the population, the large number of fishermen and the growing external pressure from industrialised fishing states. As many of the stocks in the Indian Ocean are already over-fished, any further decline in the fisheries resources could affect the human security in the area. The human impact on the Indian Ocean environment also includes pollution and degradation of marine and coastal ecosystems, both of which could affect food security in the region by damaging valuable nursery, feeding and fishing grounds.

The sea could also adversely affect humanity, as shown during the 26 December tsunami which resulted in the lost of nearly 200,000 lives and widespread economic damage. Because of this situation, there is a need to ensure that the integrity of the Indian Ocean environment is maintained. Doing so would ultimately ensure the human security of the Indian Ocean community.

Endnotes

1) Definitions of Environmental Security. http://www.acunu.org/millennium/es-2def.html Accessed on 30 June 2005.

2) Chowdhury, N. 2001 Managing the Indian Ocean Fisheries: A Collective Responsibility. Inaugural Address at the Forging Unity: Coastal Communities and the Indian Ocean's Future Conference organized by the International collective in Support of Fishworkers and the International Oceans Institute, India. Chennai, 9 - 13 October 2001.

3) UNEP, 1989. A Coast in Common: An Introduction to the Eastern African Action Plan. UNEP, Nairobi; Payet, R.A et al, 2004. Global International Waters Assessment: Indian Ocean Islands. GIWA Regional Assessment 45b. UNEP, Nairobi.

4) Chowdhury, N Ibid.

5) FAO, 2003. State of World Fisheries and Aquaculture 2002. FAO, Rome.

6) This was highlighted recently in the BBC World documentary titled Silent Shores.

7) Promoting responsible fishing in the South West Indian Ocean. http://www.sidsnet.org/latestarc/coastal-newswire/msg00058.html Accesses on 28 June 2005.

8) Hot Pursuit of Toothfish Pirates. http://news.bbc.co.uk/2/low/asia-pacific/3159229.stm Accessed on 1 July 2005.

9) Payet et al. 2004. Ibid. p.38.

10) Akimoto, K, 2001. The Current State of Maritime Security: Structural Weakness and Threats in the Sea Lanes. Paper presented at the Conference on Maritime Security in Southeast Asia and Southwest Asia organized by the Institute for International Policy Studies. Tokyo 11 - 13 December, 2001.

11) UNEP, 1989. Ibid. p.29

12) Largest Oil Spills. http://www.scu.edu.au/schools/edu/student_pages/1999/tryfell/OIL%20WEB/Largest%20Spills/Table%202.html Accessed on 27 June 2005.

13) Payet et al. ibid p.33

14) Integrated Coastal Management Country Profile: India. http://www.globaloceans.org/country/india/india.html Accessed on 20 June 2005.

15) Groombridge, B and Jenkins, M.D. 2002. World Atlas of Biodiversity. Prepared by the UNEP World Conservation Monitoring Centre. University of California Press, Berkeley, USA.

16) UNEP, 1989. Ibid. p.5;

17) Bryant, D et al. 1998. Reefs at Risk: A Map-Based Indicator of Threats to the World's Coral Reefs. World Resources Institute, Washington D.C.

18) Goombridge and Jenkins, 2002. Ibid p. 126.

19 Payet et al, 2004. Ibid. p.37.

20) Bryant et al. Ibid. p. 21

21) UNEP, 1989. Ibid p. 14.

22) FAO Assessing Damage in Countries Affected by Tsunami in South Asia. http://www.fao.org/newsroom/en/news/2004/56521/index.html. Accessed on 1 July 2005.

23) Fishermen Suffer Huge Material Toll from Tsunami, UN Figures Show. http://www.un.org/apps/news/story.asp?NewsID=13026&Cr=tsunami&Cr1. Accessed on 1 July 2005.

24) Indian Ocean Tsunami Alert Deal. http://news.bbc.co.uk/2/hi/south_asia/4639727.stm. Accessed on 30 June 2005.

Chapter 13

Securing the Seas as a Medium of Transportation in Southeast Asia

Joshua Ho

The Sea as a Medium of Transportation

Even before the term sea lanes of communications (SLOCs) was coined, the sea had been used as a medium for the exchange of goods, news and ideas (Till, 2004, pp. 8-10). According to Geoffrey Till, centuries ago, the Makassar peoples made annual pilgrimages to the northern Australian waters in search of sea cucumbers to trade with the Chinese and in the process developed trade relations with the first Australians. Similarly, fishermen from different areas in Europe met each other at sea as they followed the shoals of migratory fish far out from shore. The same patterns developed in the Asia-Pacific, as well as in the Indian Ocean, the Arabian Sea and the Mediterranean, resulting in regional, overlapping sea-based trading communities that interacted at key nodal points. The maritime trade routes developed into a complex web of inter-regional, regional, and sub-regional maritime transportation systems that spanned the globe, and the sea turned the world into a complex maritime system based on international trade. These routes became the SLOCs of today.

The main purpose of this chapter is to describe the increasing use of the major SLOCs in Southeast Asia and then to consider some of the principal security threats D in particular, piracy and maritime terrorism - to this maritime trade. The chapter will then move to a consideration of the various individual, bilateral and multilateral initiatives that have been adopted in order to deal with these threats. It is argued that, since regional states share significant maritime interests, the topic of maritime security needs to remain high on the regional agenda if we want to create a stable maritime environment. The creation of a stable maritime environment, in turn, will advance the building of the ASEAN Security Community.

Major Sea Lanes in Southeast Asia

The major sea lanes in southeast Asia are constricted at key straits such as the Malacca and Singapore Straits, the Sunda Straits and the Lombok Straits. The Straits of Malacca is 600 miles long, and is the main corridor between the Indian Ocean and the South China Sea. The major sea lanes used by tankers from the

Middle East are the Straits of Malacca and Singapore and about 26 tankers, including three fully-loaded supertankers heading for Asian ports, pass through the Strait daily. Because the Strait is relatively shallow, being only 23 metres deep at some points, an under keel clearance of 3.5 meters maximum draught is mandated by the International Maritime Organisation for transiting ships. The navigable channel at its narrowest point is only 1.5 miles wide. In terms of total volume, more than 200 boats pass through the Straits of Malacca on a daily basis, or about 60,000 on an annual basis, carrying 80 percent of the oil transported to northeast Asia (Brandon, 2000). In terms of value, the total tonnage carried by the Malacca Straits amounts to 525 million metric tonnes worth a total of US$390 billion (Kawamaura, 1998, pp. 15). The amount of traffic makes it the busiest Straits in the world currently and it is likely to be even busier in the future as a result of increasing trade flows and energy demands in Asia. According to Lloyd's List bulletin, new orders for 200 LNG carriers will be required to satisfy the growth in demand during the next 15 years. The trend of increasing traffic has also been observed for the traffic data reported via STRAITREP from 1999 to 2003, which indicate that traffic in the Malacca Straits has increased by 42 percent within the five-year period.

The Lombok Strait is wider, deeper and less congested than the Strait of Malacca. The minimum passage width in the Lombok Straits is 11.5 miles and the depths are greater than 150 metres. It is therefore considered the safest route for supertankers and the larger of these eastbound ships sometimes transit this channel. For example, tankers that exceed 200,000 deadweight tonnes (DWT) have to divert through the Lombok Straits due to the depth constraints of the Malacca Straits. Most ships transiting the Lombok Straits also pass through the Makassar Straits, which has an available width of 11 miles and a length of 600 miles. About 3,900 ships transit the Lombok Straits annually, and in terms of value, the total tonnage carried by the Lombok Straits amounts to 140 million metric tonnes, worth a total of US$40 billion (Kawamura, 1998). Ships carrying iron ore from Australia to China also enter the Indonesian archipelago through the Lombok Strait.

The least of the three straits is the Sunda Straits. It is 50 miles long and is another alternative to the Malacca Straits. Its northeastern entrance is 15 miles wide, but because of its strong currents and limited depth, deep draught ships of over 100,000 deadweight tonnes do not transit the strait and it is not heavily used. About 3,500 ships transit the Sunda Straits annually and in terms of value, the total tonnage carried by the Sunda Straits is 15 million metric tonnes worth a total of US$5 billion (Kawamura, 1998).

In addition to the transportation of oil and iron ore to the major economies in

Northeast Asia like China, Japan, Taiwan and South Korea, the Malacca Straits and the Sunda Straits also carry a significant amount of container traffic given that large ports sit astride both these sea lanes. The ports that lie along the Malacca and Singapore Straits include Singapore, Port Klang, and Tanjung Pelepas. The fourth port, Tanjung Priok, sits astride the Sunda Straits. In addition, Singapore is a major transhipment hub and sits astride the main east-west route within the global hub-and-spoke container network. To give an idea of how much container traffic was handled at each port, based on 2004 data, Singapore was the 2nd largest container port in the world, handling 20.6 million TEUs; Port Klang was the 13th largest container port in the world, handling 5.2 million TEUs; Tanjung Pelepas was the 16th largest container port in the world, handling 4 million TEUs, and Tanjung Priok was the 23rd largest container port in the world, handling 3.3 million TEUs (Boyes and Degerlund, 2005, pp. 77).

Because the Malacca, Lombok and Sunda Straits are so important to the transportation of oil and raw materials such as iron ore, as well as for the conveyance of container traffic, the free and safe navigation of commercial vessels in these sea lanes become important issues. Indeed, the importance of the regional sea lanes will increase in the future given the heavy regional dependence on the sea as an avenue for trade and transportation of energy and raw materials and the rising economic trajectory of East Asian states. In this respect, piracy and terrorism are threats to the security of shipping in the sea lanes of Southeast Asia.

Piracy

According to the International Chamber of Commerce's International Maritime Bureau (IMB), the number of piracy attacks on shipping throughout the world in 2004 was 325 (ICC IMB, 2005, pp. 4-11). This represents a significant drop in the number of attacks from the previous year of 445 in 2003, but is still the fifth highest rate since data were collected in 1992. The highest number of incidents of piracy occurred in 2000 when 469 incidents were reported. Despite the drop in worldwide pirate attacks, attacks in the Malacca and Singapore Straits continued unabated and in fact increased by 50 percent from 30 to 45 incidents. This is the second highest number of attacks in the Malacca and Singapore Straits since the IMB Piracy Reporting Centre commenced compiling statistics in 1991. Attacks in Indonesia continued to be high, accounting for 29 percent of incidents worldwide as compared to 27 percent worldwide in 2003. However, it must be mentioned that the figures for 2005 have shown a drop over that of 2004, with attacks in the Malacca Straits accounting for only one-third of the figure in 2004. This is probably a result of the increased policing efforts by the littoral states over the past year. Part of the decrease could also have been attributed to the December 2004 tsunami that devastated most of the northern coast of Sumatra, an area which is known for its piracy.

Having said that, the attacks continue to be lethal as incidents involving pirates armed with guns continue to be high at 27 percent in comparison to 23 percent in 2003. Violence continues to rise and the number of crew killed increased to 30 from 21 in 2003. Hijacking of tugs and barges and kidnapping crew for ransom continues to increase, especially in Indonesian waters and in the Northern Malacca Straits (ICC, IMB, 2005, p. 16). These attacks were initially believed to be the work of the GAM (Gerakan Aceh Merdeka), or the Free Aceh Movement, as a means to finance its activities. However, it appears that criminal syndicates are becoming increasingly involved in operating from fishing boats and conducting kidnappings as an easy way to make money.

It is reported that three types of groups typically perpetrate sea piracy in Southeast Asia: (1) small criminals, (2) well-organised criminal gangs, and it is said, (3) armed separatists (Chalk, 1997). Although piracy has been an ongoing activity in the region for a long time, what makes piracy dangerous now is that these gangs appear to be better equipped and organised than most naval authorities and have demonstrated an increased propensity to use violence. They make use of speedboats, modems, radar, satellite phones, VHF radios, and modern weaponry to take control of merchant ships. They also use hijacked ships for human smuggling and for the transport of illicit drugs and weapons (Carpenter and Wiencek, 2000, pp. 92-3). Crime syndicates involved in piracy incidents take advantage of governments that lack the financial resources, political will and efficient law enforcement agencies to tackle their criminal activities.

There are reports of up to 5 criminal syndicate groups operating in the Malacca Straits alone. The high piracy rates and its lethality have driven up shipping costs through higher insurance rates added to a number of cargoes. Estimates of the cost of pirate attacks have put these at around US$250 million a year (Farnham, 2001). Not many ship owners have adopted measures to combat piracy, probably due to the prohibitive costs involved. Already for the anti-terrorism measures, the Organisation for Economic Cooperation and Development has estimated that the costs required to implement the slew of initiatives that has been developed since 9/11 will require an initial capital cost of at least US$1.3 billion to ship operators and will incur additional annual operating costs of around US$730 million (OECD, 2005).

The emphasis on combating piracy is important, as sea piracy has been linked to the threat of maritime terrorist attacks since the events of 11 September 2001.

Young and Valencia note that "the conflation of "piracy" and "terrorism" has become common in the mass media and in government policy statements," although they themselves do not agree with this conflation (Young and Valencia, 2003). It has also been recognised that the motivations of the terrorist and those of the pirates are fundamentally different. However, we must continue to watch for the possibility of an overlap between piracy and maritime terrorism simply because the manner of operations is similar and it is difficult to distinguish between the two when an incident is unfolding. Piracy thus forms the background noise from which maritime terrorist attacks may materialise.

Maritime Terrorism

Another threat to resource and trade security is the spectre of maritime terrorism. In the new era of globalisation, ports have evolved from being traditional interfaces between sea and land to providers of complete logistics networks brought about chiefly by containerisation. Containerisation has made it possible for the carriers to shift from a port-to-port focus to a door-to-door focus. This is due to the interchangeability of the various modes of transporting the container (by road, rail, or sea), also know as intermodalism, whereby it has become possible for goods to move from the point of production, without being opened, until they reach the point of sale or final destination. Ports are also being differentiated by their ability to handle the latest generation of container ships coming on stream. According to a study by Ocean Shipping Consultants, for example, it is expected that by 2010, 8,000 TEU ships will be dominant in all trades. Concepts for a containership of 18,000 TEUs, the draught of which will maximise the available depth of the Malacca Straits, are already on the drawing board (Coulter, 2002). Hence, the dual trend of ports having to be providers of complete logistics networks and being able to handle large containerships coming on line mean that high-volume, mainline trade will focus on just a few mega-ports, making these ports the critical nodes of global seaborne trade (Flynn, 2002).

So important are hub ports in the global trading system that it has been estimated that the global economic impact from a closure of the hub port of Singapore alone could easily exceed US$200 billion per year from disruptions to inventory and production cycles. The shutting down of the ports in the western coast of the US in October 2002 due to industrial action had cost the US up to a billion dollars a day and also highlights the importance of hub ports as crucial nodes in world trade (Bush, 2002).

Hub ports therefore are potentially lucrative targets for the terrorist. Maritime terrorists may hijack carriers of liquefied petroleum gas and turn them into floating bombs to disable ports (Richardson, 2003). For example, the destruction that can be caused by such floating bombs is severe, as the detonation of a tanker carrying 600 tonnes of liquefied petroleum gas would cause a fireball of 1,200 metres in diameter, destroying almost everything physical and living within this range. Beyond this range, a large number of fatalities and casualties would occur (Sheppard, 2003). Other possible scenarios for maritime terrorism include the detonation of a 'dirty bomb' in a hub port. The 'dirty bomb' is a conventional bomb configured to disperse radioactive material and could be smuggled through a container in a container ship (Richardson, 2004, pp.112-114).

Besides attacks on hub ports, attacks on shipping can also be an attractive option for maritime terrorists. If attacks on shipping become severe, it is possible that ships may choose to divert from the current sea lanes to a safer route. The diversion could also impose costs to industry. A study undertaken by the US National Defense University has concluded that if the Malacca, Sunda, Lombok, Makassar Straits and the South China Sea were blocked, the extra steaming costs would account for US$8 billion a year based on 1993 trade flows (Coulter, 2002, pp. 139). No doubt, the cost will be even higher if current trade flows were used for the cost estimate.

In addition, prominent officials have also indicated that commercial shipping could be potential targets. For example, at the 2003 Shangri-La Dialogue, Singapore's Deputy Prime Minister (DPM), Dr Tony Tan, had warned that with the hardening of land and aviation targets, the threat of terrorism is likely to shift to maritime targets, particularly commercial shipping (Tan, 2003). Besides Dr. Tan, other officials have also warned of the possibility of maritime terrorism. For example, on 5 August, England's First Sea Lord and Chief of the Naval Staff, Admiral Sir Alan West, had warned that al-Qa'eda and other terrorist groups were plotting to launch attacks on merchant shipping. He also said that seaborne terrorism could potentially cripple global trade and have grave knock-on effects for developed economies (Lloyd's List, 2004a). Besides statements by prominent officials, there are also possible indications that Southeast Asian terrorist groups may have begun to look at the maritime domain as a new avenue for attacks.

However, one of the most definitive statements that local terrorist groups have been setting their sights on commercial shipping came from Indonesia's National Intelligence Agency, which revealed that detained members of Southeast Asian Islamic terror group Jemaah Islamiah, which is linked to al-Qa'eda, admitted that shipping in the Malacca Strait had been a possible target (Lloyd's List 2004). The discovery of plans detailing vulnerabilities in US naval fleets on al-Qa'eda linking terrorist suspect, Babar Ahmad, also places beyond a shadow of doubt that al-Qa'eda terrorist groups have been looking at the maritime domain as a possible mode of attack (Asia Times, 2004).

Individual Counter-Measures

Having detailed the nature of the threats of piracy and maritime terrorism, it must be said that the regional countries are already taking steps to address the issues. For example, according to its former Chief of Naval Staff, the Indonesia Navy is responding to the increasing trend of piracy in its waters by promoting a package of reforms and modernising the Navy's platforms towards a new emphasis on coastal interdiction and increasing patrols against illegal activities in their own waters (Karniol, 2003). Indonesia has also formed Navy Control Command Centres (Puskodal) in Batam and Belawan with equipment and the placement of special forces which can respond to armed hijackings and piracy (Sondakh, 2004). The Indonesian Chief of Naval Staff has urged the shipping community to contact the two Control Command Centres if it faces problems with piracy in Indonesian waters.

In addition to the hard measures adopted, the Indonesian Ministry of Home Affairs has also undertaken dissuasion programmes. These programmes focus on the alleviation of poverty and the increase of the people's welfare in the remote areas. In particular, the six Regencies of Rokan, Hilir, Bengkalis, Siak, Palawan, Indragiri Ilir and Karimun, that border the Malacca and Singapore Straits, are currently the main priority areas. The next priority is then given to the tens of Regencies that border the other SLOCs through Indonesia (Magindaan, 2004).

Malaysia has also taken action to keep the piracy rates low in the Malacca and Singapore Straits. For example, the Royal Malaysian Navy has built a string of radar tracking stations along the Straits of Malacca to monitor traffic and has acquired new patrol boats largely to combat piracy (Brown, 2003). At the maritime enforcement level, a special anti-piracy task force was established by the Royal Malaysian Marine Police in 2000 with immediate acquisition of 20 fast strike craft and 4 rigid inflatable boats (RIBs) at a cost of RM15 million. Recently, 60 marine police officers are also being trained as a marine police tactical unit (commando). This unit will be assisted by another two elite forces in the Police Department Ð Special Action Forces and 69 Commando Unit, and they will accompany the marine police units. The unit will be deployed along the Straits of Malacca (Sazlan, 2004). In addition, the Malaysian Police will also deploy assault weapons on tugs and barges plying the busy shipping lanes of the Malacca Strait in response to two attacks involving tugs in March 2005 after a long absence of piracy in Malaysian waters (Lloyd's List, 2005). The Royal Malaysian Navy has also intensified its training activities and patrols in the northern reaches of the Malacca Straits beyond the area of the one fathom bank in an effort to increase presence and thus deter both piracy and maritime terrorism (Nor, 2005).

Another important measure adopted by the Malaysian government is the formation of the Malaysian Maritime Enforcement Agency (MMEA), the equivalent of a coast guard, that was set up in November 2005. The MMEA will bring together several existing maritime enforcement agencies such as the Royal Malaysian Marine Police, the Fisheries Department, Immigration Department, Customs Department, and the Marine Department. The consolidation of maritime-related agencies into a single command of the MMEA will enable more focus and enhance ability to deal with maritime-related offences (Sazlan, 2004). The MMEA will also be involved in enforcement duties and search and rescue.

Singapore has also implemented a range of measures to step up maritime security. These include an integrated surveillance and information network for tracking and investigating suspicious movements; intensified navy and coastguard patrols; random escorts of high-value merchant vessels plying the Singapore Straits and adjacent waters; and the re-designation of shipping routes to minimise the convergence of small craft with high-risk merchant vessels (Scott, 2003). In addition to increasing its own patrolling activities, Singapore has also cooperated closely with the International Maritime Organisation (IMO) by implementing amendments to the International Convention for the Safety of Life at Sea in the form of the International Ships and Port Facility Security (ISPS) Code, which came into effect in July 2004. Singapore has also signed the 1988 Rome Convention on the Suppression of Unlawful Acts against the Safety of Maritime Navigation (SUA Convention). The Convention would extend the rights of maritime forces to pursue terrorists, pirates and maritime criminals into foreign territorial waters and provides guidelines for the extradition and prosecution of maritime criminals. Singapore will also be erecting radiation detectors at its ports to scan containers for nuclear and radioactive material under the US Mega-ports Initiative (Boey, 2005). The Republic of Singapore Navy has also formed the Accompanying Sea Security Teams (ASSeT), similar to armed marshals, to board selected merchant ships proceeding into and out of harbour to prevent the possibility of a ship being taken over by terrorists (Lian, 2005).

Bilateral Counter-Measures

Besides individual measures, there have been efforts at bilateral cooperation based on a web approach. Indonesia and Singapore agreed in 1992 to establish the Indonesia-Singapore Coordinated Patrols in the Singapore Straits. This has involved the setting up of direct communication links between their navies and the organisation of coordinated patrols every three months in the Singapore Straits (Go, 2002). Singapore and Indonesia have also set up a Joint Radar Surveillance system that will monitor traffic in the Straits, and that will also provide the position, course and speed of the shipping in the Straits (MINDEF, 2005). Indonesia and Malaysia decided in 1992 to establish a Maritime Operation Planning Team to coordinate patrols in the Straits of Malacca. The Malaysia-Indonesia Coordinated Patrols are carried out four times a year, and so is the Malaysia-Indonesia Maritime Operational Coordinated Patrol, which is conducted together with other maritime institutions, like customs, search and rescue and police from the two countries (Sondakh, 2004).

Besides the three littoral states, other countries are also beginning to get involved in the security of the Malacca Straits. For example, India has begun talks with Indonesia on how to improve maritime security in the northern part of the Malacca Straits. Thailand has recently expressed interest in contributing to the security of the Malacca Straits, especially in terms of capacity building. However, in both of these cases concrete measures have yet to materialise. China has also recently signed a strategic partnership agreement with Indonesia and one of the items is increased maritime cooperation that could include joint efforts to combat smuggling and piracy (International Herald Tribune, 2005). At a more concrete level, the US has conducted anti-piracy exercises with Indonesia which have involved the boarding and inspection of shipping. This exercise was referred to as Crisis Action Planning SMEE 05-03 (Gatra, 2005).

Multilateral Counter-Measures

In comparison to the bilateral cooperation that exists in Southeast Asia, the multilateral response to piracy and terrorism has been more limited and is only starting to take shape. Although many multilateral forums exist, like the Asia-Pacific Economic Cooperation (APEC) grouping, the Association of Southeast Asian Nations (ASEAN), the ASEAN Regional Forum (ARF), and ASEAN Plus Three, very few concrete measures have actually materialised from these high level forums. The few operational measures that have materialised arise from the ARF and the ASEAN Plus Three framework.

The ASEAN Regional Forum (ARF)
The ASEAN Regional Forum (ARF) currently comprises 24 countries, namely the ten ASEAN countries, Australia, Canada, China, European Union, India, Japan, DPRK (North Korea), South Korea, Mongolia, New Zealand, Pakistan, Papua New Guinea, Russian Federation and the United States.

The ARF adopted the Statement on Cooperation against Piracy and other Threats to Maritime Security at the 10th ARF Post Ministerial Conference held in Cambodia in June 2003. In this document, ARF participants regarded maritime security as "an indispensable and fundamental condition for the welfare and economic security of the ARF region". The ARF participants also expressed their commitment to becoming parties to the Convention for the Suppression of Unlawful Acts against the Safety of Maritime Navigation (SUA) Convention and its protocol. The SUA protocol extends coastal state enforcement jurisdiction beyond the territorial limits, and, in particular circumstances, allows the exercise of such jurisdiction in an adjacent state's territorial sea. It also allows the state to prosecute criminals for crimes committed in another state's territorial waters. To date, half of ASEAN members have signed the convention, namely Brunei, Myanmar, Singapore, Philippines, and Vietnam. Malaysia has also indicated that they would sign the convention sometime in the near future.

ASEAN Plus Three

The ASEAN Plus Three forum comprises the ASEAN states together with the countries of China, Japan, and South Korea. The ASEAN Plus Three is an attempt to build a regional association that is more limited in its geographic membership than APEC or the ARF. In November 2001, at the ASEAN+3 Summit in Brunei, Japanese PM Koizumi proposed the convening of a governmental-level working group to study the formulation of a regional cooperation agreement related to anti-piracy measures. Acceptance of this proposal has led to negotiations for the establishment of the "Regional Cooperation Agreement on Combating Piracy and Armed Robbery against Ships in Asia" (ReCAAP) among representatives of the ASEAN states, China, Japan, South Korea, India, Sri Lanka and Bangladesh. At a meeting on the 11 November 2004 in Tokyo, the 16 states agreed to the setting up of an Information Sharing Centre (ISC) in Singapore (The Straits Times Interactive, 2004). The Information Sharing Centre will have a full-time multinational staff to maintain a database for piracy-related information and facilitate communication among national agencies prosecuting piracy cases. The ten participating states of ReCAAP will have to ratify the agreement before the Information Sharing Centre can be set up. So far, eight countries - Brunei, Cambodia, Japan, Laos, Myanmar, Philippines, Singapore and Thailand - have signed the initiative. It is expected that India and South Korea will be signing the agreement soon and both Indonesia and Malaysia have expressed support for the agreement. The setting up of the Information Sharing Centre is important as it will bring critical analysis to bear on the whole topic of piracy, based on information made available through the government agencies. More ReCAAP participating countries should accede to the formation of the ISC and do it soon.

Other Multilateral Arrangements

Besides the agreements and the arrangements that arise out of the existing multilateral mechanisms, there are three other arrangements which have not originated from these more formal mechanisms, but are nevertheless important. The three arrangements include the Five Power Defence Agreement, the Western Pacific Naval Symposium, and the controversial Regional Maritime Security Initiative, which resulted in the three littoral countries coming together to conduct the Malacca Straits Coordinated Patrols.

The Five Power Defence Agreement (FPDA)

The FPDA was founded in 1971 and brings together Australia, Malaysia, New Zealand, Singapore, and the United Kingdom in a consultative defence arrangement. The FPDA was formed primarily as a postcolonial response to the Indonesian Confrontation as the members have agreed to consult each other should there be a threat to the security of each of the member countries. Its original focus on conventional threats has now given way to more non-conventional threat scenarios. Recently, the FPDA agreed to expand the scope of its activities to include non-conventional security threats, such as maritime terrorism. The FPDA has also conducted an anti-terror drill as part of Bersama Lima in September 2004 (Boey, 2004).

Western Pacific Naval Symposium (WPNS)

The WPNS was created in 1988 and brings together 18 member navies - Australia, Brunei, Cambodia, China, France, Indonesia, Japan, Malaysia, New Zealand, Papua New Guinea, Philippines, Russia, Singapore, South Korea, Thailand, Tonga, United States and Vietnam, plus the 3 observer navies of Canada, Chile, and India (WPNS web site). The administrator of the WPNS is the US Pacific Command. The WPNS was originally a forum used to promote mutual understanding among navies of the region and aims to increase naval cooperation in the Western Pacific among Navies by providing a forum for the discussion of maritime issues, both global and regional, and in the process, generate a flow of information and opinion among naval professionals leading to common understanding and possibly agreement. The WPNS has now grown to include regular shore-based and sea exercises. It was decided recently that coast guard agencies will be invited to participate in the next WPNS sea exercise and it is hoped that this will enhance inter-agency coordination and understanding (Tay, 2005). Another new initiative is known as the 'Connecting Networks for the Enhancement of Knowledge Sharing.' This initiative aims to allow non-navy agencies and inter-governmental agencies to be invited to present relevant topics of interest at workshops and symposia. The WPNS has recently concluded a multilateral sea exercise in May

2005 of which one of the aims was to improve the inter-operability among participating navies through the compilation of the sea situation picture and the sharing of data through a common data link. Participation in WPNS activities is voluntary.

The Regional Maritime Security Initiative (RMSI)

The Regional Maritime Security Initiative (RMSI) is very much the brainchild of the U.S. Pacific Command and initiated perhaps due to the frustration at the slowness in the implementation of concrete measures in the region to tackle the transnational maritime terrorist threat (US Pacific Command web site). The RMSI is a long-term, multi-national approach to counter transnational threats, including terrorism, maritime piracy, illegal trafficking, and other criminal activities in the maritime domain. RMSI intends to be a partnership of regional states that are willing to contribute their resources to enhance maritime security. It is not a treaty or an alliance and will not result in a standing naval force patrolling the Pacific. The goal of the RMSI is "to develop a partnership of willing regional nations with varying capabilities and capacities, to identify, monitor, and intercept transnational maritime threats under existing international and domestic laws." The RMSI aims to build and synchronise inter-agency and international capacity, to harness available and emerging technologies, to develop a maritime situational awareness to match the picture that is available for international airspace, and to develop responsive decision-making structures that can call on immediately available maritime forces to act when required. Despite its laudable goals, it appears that the RMSI is still very much in the preliminary planning phase, and that PACOM may now be using the Western Pacific Naval Symposium (WPNS) as a forum to advance the concept of the RMSI.

Operation MALSINDO (Malacca Straits Coordinated Patrols)

When the RMSI was announced by PACOM, the littoral states of Malaysia and Indonesia perceived it as a means by which the US would conduct operational patrols in the Malacca Straits to secure its own maritime self-interest. Both were opposed to the notion of patrols conducted by extra-regional states, whilst Singapore was more open to its conduct. As a by-product of the RMSI, and in response to the concerns expressed by the US over the security of vessels transiting the Malacca Straits, Operation MALSINDO was born (Vijayan, 2004). Currently, 17 ships have been allocated to the patrols, seven from Indonesia, five from Malaysia, and five from Singapore. The first trilateral naval patrols were launched in July 2004 aimed at reducing piracy and smuggling activities in the Straits on a 24/7 basis. Each navy only patrols within the territorial waters of its respective country. To be more effective, it may be necessary to explore the possibility of conducting Joint patrols where resources are pooled for the common task.

Securing the Seas

In conclusion, the rise of Asia, principally driven by the economies of China, India and Japan, will result in increases in inter and intra-regional trade flows, as well as increases in energy demand, both of which mean an increasing reliance on the sea as a mode of transport. This surge in the use of the sea as a mode of transport implies that the security and the safeguarding of the sea lanes become more crucial than ever. Hence, besides individual measures, there is a need to move towards a more cooperative regime between both the littoral states as well as other stakeholders to enhance the security of the sea lanes as the threats are transnational in nature. An act of armed robbery that occurred at the end of February 2005 demonstrated the transnational character of the threat to shipping in the sea lanes. The incident involving a Japanese tug occurred in Malaysian waters, and the Japanese crew were taken as hostages. The perpetrators from Indonesia were suspected to be responsible for the incident. The hostages were finally released in the vicinity of Southern Thailand after the Japanese owners paid the ransom.

As countries in the region share significant maritime interests, the topic of maritime security needs to remain high on the regional agenda in order to create and maintain a stable maritime environment. The creation of a stable maritime environment in turn will advance the building of the ASEAN Security Community, since the declaration of the Bali Concord II states that "maritime cooperation between the ASEAN member countries will contribute to the evolution of this Security Community" (ASEAN Secretariat, 2004). However, one important aspect to keep in mind as we pursue the creation of this Security Community is that the littoral states will have the primary responsibility in addressing maritime security issues with important roles for other stakeholders, and that any new initiative which is proposed should be in accordance with the rule of international law (Boey and Chin Lian, 2005).

References

ASEAN Secretariat (2004), 'Declaration of ASEAN Concord II (Bali Concord II)'. Available at http://www.aseansec.org/15159.htm <Accessed on 27 June 2005>.

Asia Times (2004), 'Terror on the High Seas', 21 October.

Boey, D. (2004), 'FPDA tackles terror threat in drill; The five defence partners add a new facet to exercise, and will hunt down and board a 'hijacked ship' in the South China Sea', *The Straits Times Interactive,* 11 September.

Boey, D. (2005), 'Radiation detectors for Singapore port', *The Straits Times Interactive*, 11 March.

Boey, D. and Goh Chin Lian (2005), 'ARF states should stage joint drills; Defence Minister asks Asean Regional Forum to go beyond talks to boost maritime security', *The Straits Time Interactive*, 3 March.

Boyes, J. R. C. and Jane Degerlund (2005), 'Rising to the top', *Containerisation International*, March.

Brandon, J. J. (2000), 'Piracy on High Seas is Big Business', *International Herald Tribune*, 28 December.

Brown, N. (2003), 'Malaysia asks for Help to Fight Piracy', *Jane's Navy International*, 1 November.

Carpenter, W. M. and David G. Wiencek (2000), 'Maritime Piracy in Asia', in W. Carpenter and D. Wiencek (eds.), *Asian Security Handbook 2000* (Armonk, New York: M.E. Sharpe), pp. 92-93.

Chalk, P. (1997), *Grey-Area Phenomena in Southeast Asia: Piracy, Drug Trafficking and Political Terrorism*, (Canberra: Canberra Papers on Strategy and Defence No. 123, Strategic and Defence Studies Centre, Australian National University, 1997), Chapter 2.

Coulter, D. Y. (2002), 'Globalisation of Maritime Commerce: The Rise of Hub Ports', in Sam J. Tangredi (ed), *Globalization and Maritime Power* (Washington D.C.: National Defense University Press, December), pp. 135-138.

Declaration of ASEAN Concord II

Farnham, A. (2001), 'Pirates!', *Fortune,* 15 July.

Flynn, S. E. (2002), 'America the Vulnerable', *Foreign Affairs,* January/February, pp. 60-74.

Gatra (2005), 'TNI starts anti-piracy exercise with US military', *Gatra,* 2 May 2005. Available at http://www.gatra.com/2005-05-02/artikel.php?id=84037 <Accessed on 9 May 2005>

Go, R. (2002), 'Singapore Strait Patrols Keep Pirates at Bay', *The Straits Times* Interactive, 16 May.

Goh Chin Lian (2005), 'Armed Navy escorts for suspect ships', *The Straits Times Interactive*, 28 February.

Lloyd's List (2004a), 'First Sea Lord warns of al-Qa'eda plot to target merchant ships", 5 August.

Lloyd's List (2004b), 'Malacca Strait is terror target admit militants', 26 August.

MINDEF (2005), 'Singapore and Indonesian Navies Launch Sea Surveillance SystemÕ, *News Release*, 27 May. Available at http://app.sprinter.gov.sg/data/pr/20050527997.htm <Accessed on 2 June 2005>

ICC International Maritime Bureau (2005), 'Piracy and Armed Robbery against Ships Annual Report: 1 January Ð 31 December 2004', January.

International Herald Tribune (2005), 'China and Indonesia seal strategic pact', 26 April. Available at http://www.iht.com/articles/2005/04/25/news/indonesia.php <Accessed on 9 May 2005>

Karniol, R. (2003), 'Indonesian Navy to Focus on Coastal Interdiction', *JaneÕs Defence Weekly*, 12 November.

Kawamura, Sumihiko (1998), 'Shipping and Regional Trade: Regional Security Interests', Sam Bateman and Stephen Bates, eds., *Shipping and Regional Security*, Canberra: Strategic and Defence Studies Centre, The Australian National University.

Magindaan, R. (2004), 'Maritime Terrorism Threat: An Indonesian Perspective', *Paper presented at Observer Research Foundation Workshop on Maritime* Counter Terrorism, 29-30 November.

Nor, Adm Dato' Sri Mohd Anwar bin HJ Mohd, Chief of Navy, Royal Malaysian Navy (2005), 'Malaysia's Approach', *Presentation at ARF Regional Cooperation in Maritime Security Conference*, 2-4 March.

OECD (2005), Report of the Organisation for Economic Cooperation and Development, 'Price of Increased Maritime Security is Much Lower than Potential Cost of a Major Terror Attack', at: http://www.oecd.org/document /30/0,2340, en_2649_201185_4390494_1_1_1_1,00.html <Accessed on 5 April 2005>

Richardson, M. (2003), 'Terror at Sea: The World's Lifelines are at Risk', *The Straits Times Interactive*, 17 November. Available at http://straitstimes.asia1.com.sg/

Richardson, M. (2004), *A Time Bomb for Global Trade: Maritime Related Terrorism in an Age of Weapons of Mass Destruction*, Singapore: Institute of Southeast Asian Studies, pp. 112-114.

Sazlan, Iskander (2004), 'Counter Maritime Terrorism: Malaysia's Perspective', *Paper presented at Observer Research Foundation Workshop on Maritime Counter Terrorism*, 29-30 November.

Scott, R. (2003), 'IMDEX: Singapore Stresses Counters to Maritime Terrorism', *Jane's Defence Weekly*, 1 November.

Sheppard, B. (2003), 'Maritime Security Measures', *Jane's Intelligence Review*, 1 March.

Tan, Tony (2003), Deputy Prime Minister and Minister for Defence, Remarks at the Plenary Session on 'Maritime Security after September 11th', *Second IISS Asia Security Conference*, Singapore, 30 May-1 June.

Sondakh, Admiral Bernard Kent (2004), 'National Sovereignty and Security in the Strait of Malacca', Paper delivered at *Conference on 'The Straits of Malacca: Building a Comprehensive Security Environment'*, Maritime Institute of Malaysia, Kuala Lumpur, 11-13 October.

Tay, R. (2005), 'Multi-lateral Frameworks and Exercises: Enhancing Multi-lateral Co-operation in Maritime Security', Presentation at the *ARF Regional Co-operation in Maritime Security Conference*, 2-4 March.

The Straits Times Interactive (2004), 'Asian nations band to fight piracy', *The Straits Times Interactive*, 13 November.

Till, G. (2004), *Seapower: A Guide for the Twenty-First Century*, Great Britain: Frank Cass.

Western Pacific Naval Symposium Website. Available at http://www.apan-info.net/wpns/ <Accessed on 13 May 2005>

US Pacific Command web site, 'RMSI: The Idea, The Facts', *U.S. Pacific Command Website*, 21 December 2004. Available at http://www.pacom.mil/rmsi/ <Accessed on 13 May 2005>

Vijayan, K. C. (2004), '3-nation patrols of strait launched; Year-round patrols of Malacca Straits by navies of Singapore, Indonesia, Malaysia aimed at deterring piracy and terrorism', *The Straits Times Interactive*, 21 July.

Young, A. J. and Mark J. Valencia (2003), 'Conflation of Piracy and Terrorism in Southeast Asia: Rectitude and Utility', *Contemporary Southeast Asia*, Volume 25, Number 2, August, pp. 270-274.

CHALLENGES AND PROPSECTS FOR
MARITIME SECURITY COOPERATION

Chapter 14

Security of Sea Lanes of Communication (SLOCS) through the Straits of Malacca: The Need to Secure the Northern Approaches

Mat Taib Yasin

Introduction

The Straits of Malacca is a critical SLOC linking the Indian Ocean and the South China Sea. Given its strategic international importance, where approximately one third of global trade and more than half of the world's energy transits, no major economic and military power would ignore threats to the security and safety of navigation in the Straits. Hence, the recent surge in incidents of sea robberies/piracies in the Straits of Malacca has gained unprecedented international attention. Initially, there was an implicit threat of military intervention. This was followed by the Lloyd Joint Warfare Committee classification of the Straits of Malacca as a "war risk" zone. This classification was subsequently used to justify an increase in insurance premiums for vessels transiting through the Straits designed to pressure the littoral states to ensure the security of the Straits.

As in other parts of the world, security threats in semi-enclosed waters like the Straits of Malacca are closely linked to security developments in the surrounding areas. Most of the sea robbers are likely to come from nearby shores as would the weapons, including firearms used in some of these aggressive attacks. This chapter discusses the manner in which rampant transnational criminal activities around the northern approaches to the Straits of Malacca have posed a significant threat to the security of the Straits.

The Northern Approaches to SOM

The Scenario

The northern approaches to the SOM are made up of land masses (including islands) and a water column encompassed by a rectangular region stretching from the "Golden Triangle" in northeastern Bangladesh in the west, Cambodia in the east and the Andaman Sea in the south (Figure 1).

Figure 1
The Andaman Sea Region
(FOCAL AREA OF THIS RESEARCH MARKED BY THE RED BOX)

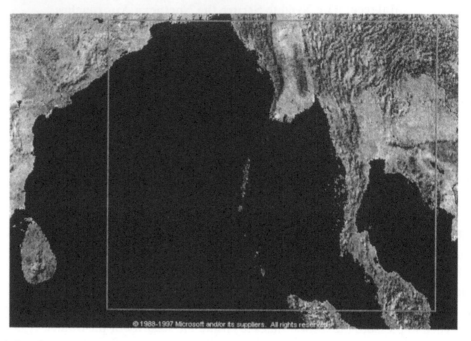

Mainly due to its geopolitical landscape, the entire region is predominantly characterised by transnational criminal activities, notably narcotics-trafficking, gun running and human smuggling. The complex clandestine arms trade draws funding from the lucrative drugs and human smuggling trades as smuggling syndicates engaged in these illicit activities import arms to protect their illegal businesses.

A major part of the drug producing "Golden Triangle" is located in Myanmar (Burma) and Thailand (Das, 2004; UN Report, 2004). Drugs produced from the region find their way to markets all over the world. With its huge profit margins, drug trafficking is by far the most lucrative means of generating funds to support the gunrunning business and fuel the ever-growing terrorist activities and insurgencies around the region. The symbiotic relationship between terrorist groups and drug cartels has provided safe and secure access for the movements of both narcotics and arms (Raman, 2002).

Firearms purchased using proceeds from the narcotics trade, apart from being merchandised commodities, are also used to protect drug trafficking operations (American Security Information Council, 1998). This has become evident in drug smuggling interception missions carried out by the authorities that have often ended with discoveries of a large cache of firearms ranging from pistols to M16 assault rifles (Salam, 2005).

Drug trafficking is also a main source of funding for insurgent movements and criminal groups around the world. The United Wa State Army (UWSA) - a splinter faction of the Burmese Communist Party operating from the northern Shan state in Myanmar - depends extensively on drug money to fuel its movement and to equip its forces.

Similarly, the notorious Sri Lankan Liberation of Tigers of Tamil Ealam (LTTE) has been known to be deeply involved in drug trafficking throughout the region. Besides transporting timber, sugar and other commercial items, their "phantom fleet" also transports narcotics from Myanmar to Turkey. They also provide protection and courier services to the seaborne drug shipments from Myanmar to various countries in Europe and the United States. The drug money is then channeled into arms purchases for continuing the Sri Lankan insurgency. Until 1995, the LTTE also used to have a base at Twante, an island off the coast of Myanmar. The Thai island of Phuket later became its main backup base. It was at this base that the LTTE was reportedly building a submersible vessel/submarine (Singh, 2003).

The LTTE has also been known to transship military hardware from Thailand, particularly Phuket and Krabi to Sri Lanka through the Andaman Sea (Asia Week, 2000). Using smaller vessels to pick up the arms consignment from remote harbours and then transferring them to bigger ships (often in mid-ocean), the arms are then ferried to Sri Lankan waters. About 60-70 nautical miles off the Sri Lankan coast, normally off the LTTE base of Mullaitivu, the consignments are then transferred to smaller boats belonging to the LTTE. These small boats are escorted to the shore by Sea Tigers.

The transfers of drugs and firearms throughout the region have been confirmed through numerous interceptions by the authorities. In 1997, the Royal Thai Navy seized a consignment of arms meant for the People's Liberation Army (Manipur) off its port of Ranong (Capie, 2002, p. 20). During two sting operations, *Operation Leech* and *Operation Poorab* in early 1998, the Indian Army uncovered arms from Laos and Thailand being transshipped through the Andaman Sea, after a huge cache of arms was seized. A number of Arakan and Karen National Union insurgents were killed in these operations. In *Operation Poorab* they intercepted two Thai vessels near Narcondan Island and seized 50 kg of heroin of Myanmar origin along with a large shipment of arms from Thailand.

According to Indian intelligence sources, up to April 1998 alone, there were as many shipments of arms from Ranong in Thailand to Cox Bazar in Bangladesh, mainly destined for Northeast Indian insurgents (The Hindustan Times, 1998). In

2004, also off Ranong, the Thai Navy intercepted a 16-metre boat after a chase and confiscated two tons of weapons and ammunition reportedly belonging to the Manipur Revolutionary People's Front. Six crew members were from the Arakan region of Myanmar. All of these arms consignment were destined for Cox Bazaar in Bangladesh.

The same routes used by gunrunners and narcotic traffickers are also used by human smuggling syndicates, an equally lucrative trade, to convey illegal immigrants from Bangladesh and Myanmar to Southeast Asia (FTU-B, 2003). Most of the Burmese illegal immigrants made their way south to Malaysia and Singapore, either direct from ports in Myanmar or by land through Thailand. Most of these illegal immigrants are sold as cheap labour to organised syndicates in Malaysia, Singapore and Thailand.

The Contributory Factors

Apart from geographical proximity, which allows for easy access to sources of drugs, firearms and cheap labour, the high demand for these illicit consignments in the relatively more developed Southeast Asia and Northeast Asia, make them a thriving trade. Firearms find a ready market among separatist groups operating along the common borders of some regional states. Among such groups are the dreaded All Tripura Tiger Force (ATTF), an underground wing of the Communist Party of India (Marxist) and the ruling government of Tripura, operating along the Bangladesh-Myanmar border (Gurung and Maitra, 1995), the Karen National Union (KNU) operating along the Myanmar-Thailand border and the Pattani United Liberation Organization (PULO) operating along the Thailand-Malaysia border.

The other factor spurring transnational criminal activities is the environment of "lawlessness and diminished security" due mainly to domestic political instability, tradition and poor law enforcement capacity, especially in areas far away from centres of administration. In this regard, most of the regional states, in varying degrees are faced with problems related to domestic security. In India, drug trafficking activities are closely linked to the country's ineffectively regulated pharmaceutical trade, especially the illicit production of methaqualone - a common synthetic drug used by drug addicts. The "hawala system" that is still practiced in the country has only facilitated narcotics money laundering.

The northern approaches to the Straits of Malacca are fringed by the Indian Andaman and Nicobar chains of islands and the Andaman Sea. Given that they are located more than 1000 km from the Indian mainland, their security may not get as much priority as regions closer the mainland. Both the Indian Navy and the Coastguard have at one time conceded that they lack the resources to effectively police all of the Andaman Sea waters (Mahapatra, 2005).

Myanmar, on the other hand, seems to lack the political will to shed its title as the world's second largest producer of illicit opium in addition to being a major source of methamphetamine and heroin. Besides being short of the necessary enforcement tools, especially at sea, to take on major narcotic-trafficking groups, a lack of serious commitment against money laundering continues to hinder the overall anti-drug effort.

Thailand shares an 8,000 km long common land border with Cambodia, China, Laos, Myanmar, and Malaysia, and common maritime borders with India, Malaysia, Cambodia and Vietnam. Given the length of its borders, the country is hard pressed to effectively police all of them and they are already known to be operating grounds for narcotic traffickers, gunrunners and human smugglers. Thailand is an illicit transit point for heroin en route to the international drug market. Also, as a producer of opium, heroin, marijuana and, more recently, amphetamine, the country is also confronted with increased indigenous abuse of the drug (Travelblog, 2005).

The porous borders also expose the country to a serious illegal immigration problem. While illegal workers from Cambodia slip across Thailand's eastern boundary, Chinese and Burmese immigrants often infiltrate through the northern borders. In the south, Ranong Province provides the easiest access to Thailand for immigrants from Myanmar. Thailand is in fact the transit hub in the region for illegal immigrants *en route* to the developed world. With the help of corrupt officials and lax penalties, the demand for service has enabled an elaborate network of human trafficking agents to flourish (The BurmaNet News, 1995).

The fact that human trafficking is now an international industry with a large global network makes addressing it a more daunting task. Easy trans-border movements of illegal immigrants could mean it is equally easy for smugglers to bring contraband like narcotics and arms into the country (The BurmaNet News, 1995). The recent unrest in the southern Provinces has added another dimension to its domestic security challenges. The chaotic environment offers a good opportunity for gunrunners and narcotic traffickers to operate. While there will be more demand for arms, the authorities already hard pressed to tackle the threats from insurgents would have less time to track and monitor the movement and trading of arms and drugs. The initial aggressive response to the threat has also inflamed bilateral tension between Thailand and neighbouring Malaysia (Pongsudhirak, 2005).

As a country that lies at the southern most tip of mainland Asia, Malaysia has allocated vast amounts of resources to guard its northern land and sea borders. Apart from manpower from various enforcement and security agencies, some parts of the border are also guarded by human-made physical barriers. Despite all of these measures, the boundaries remain too porous. Illicit firearms, drugs and illegal immigrants continue to flow into the country. The vast economic disparity between communities living on either side of the Malaysia-Thai border has turned smuggling activities into an acceptable daily way of life (Rumley, 1991). The most popular consignments are rice and diesel. However, in these smuggled rice consignments, authorities have also often found narcotics as well as firearms (Solahudin, 2005).

Lack of Inter-State Security Cooperation

No state can tackle threats of transnational criminals on its own, not even a major power. Within the region, despite the presence of ASEAN cooperative security mechanisms against non-state sponsored crimes, genuine cooperation is still lacking. On the other hand, not all regional states are members of ASEAN. Hence, there are numerous gaps in the existing cooperative mechanism. It is not too far fetched to say that the current level of cooperation among regional states is simply no match for the established regional narcotic and firearms cartels led by the likes of the Japanese Yakuza and the Chinese Triads.

Poor or non-existent inter-state cooperation could be attributed to a number of factors such as superpower geo-politicking and inter-state border geopolitics. In the course of ensuring their continuous global dominance, the United States, as the only superpower, has sought to align states to conform to its geostrategic interests. This includes practicing politics of "divide and rule". In areas where cooperation does not favour United States interests, wedges have been placed between some regional states. For example, countries like India and Pakistan have been discouraged from embarking on a joint gas pipeline project with Iran (American Foreign Policy Council, 2005). Similarly, ASEAN countries have been pressured to distance themselves from a fellow ASEAN member, Myanmar. Its military junta has been accused of having a poor human rights record. To enhance its strategy to "contain" a potentially economically and militarily powerful China, the United States has dangled attractive carrots to regional powers like India. The country was recently offered high-technology weapon systems, military R&D and access to US civil nuclear technology (Samanta, 2005).

All of this geopoliticking has, to an extent, impacted on the overall regional security context. Along with the United States seemingly aggressive approach in the war against terror, it has certainly encouraged distrust and suspicion, thereby making interstate security cooperation even more difficult.

Interstate Geo-Politicking

In general, bilateral relations between some regional states can at best be described as "occasionally irritating", mainly as a result of historical legacies (CIA web site).

India-Myanmar

Relations between India and Myanmar, for example, in the earlier days were marred by India's support for the opposition leader Aung San Suu Kyi. Of late, relations between the two countries have improved. In return for Myanmar's support for its war against Christian insurgents fighting for an independent homeland in its peripheral territories - for example, in West Bengal and Nagaland - India has now changed its posture. Their disputed boundaries were amicably renegotiated in 1967. However, a 128-kilometer Indo-Myanmar stretch from India's tri-junction with Myanmar and China is unresolved as it is linked to the Indo-China boundary dispute. The signs of improved relations between the two countries are evident in the free movement of the border population 20 kilometers on either side of the border and the construction of two roads from India to Myanmar in 2001 (Dawn, 2005).

Bangladesh-Myanmar

Bilateral relations between Bangladesh and Myanmar are occasionally strained by the issue of Muslim refugees. From 1991, about 400,000 ethnic Rohingya Muslim refugees fled from religious persecution in Myanmar and streamed into Bangladesh. Population tensions in Bangladesh have led to clashes between them and the local population. Between mid-1992 and 1997, 230,000 Rohingyas were repatriated. There is also the problem of 300,000 Bihari Muslims who are living in abject poverty in Bangladesh because they are not accepted in Pakistan.

Myanmar-Thailand

Relations are occasionally marred by insurgency activities along their common border. Karens from Burma flee into Thailand to escape fighting between Karen rebels and Burmese troops, resulting in Thailand sheltering about 118,000 Burmese refugees in 2004.

Malaysia-Thailand

Although most of the boundary issues between the two countries have been resolved, trans-border irritations do occasionally arise. The current unrest in Thailand's predominantly Muslim southern Provinces has created some tension between the two countries. The apparent ill-treatment of Thai Muslims by the Thai authorities has not been well received by Malaysians. This sympathy for Thai Muslims, who share common culture and ethnicity with their brethren across the border, on the other hand, reinforces Thailand's suspicion that separatist movements in the south have been getting unofficial support from Malaysia. Sensitivities arising from these issues often make border security cooperation against trans-boundary criminals less effective.

The Spillover into the Straits of Malacca

The flow of drugs and firearms from the criminal infested region southwards has been going on for centuries. Cambodian weapons flow through southern Thailand from ports such as Phuket and Ranong. The arms either travel by boat, usually fishing boats, across the Malacca Straits to Aceh or southwards into Malaysia. Others move by land southward into Malaysia. Some of the arms consignments even flow as far as the southern Philippines, Japan and the South Pacific. The Cambodian Prime Minister, Hun Sen, has personally admitted that the country is one of the major suppliers of arms to insurgents operating in Southeast Asia (Berita Harian, 2005).

A very similar modus operandi is used by drug traffickers. Drug consignments are smuggled by high-speed boats, normally at night. They are ferried to predetermined pick up points, usually on isolated islands like Pulau Payar and Pulau Bidan, both located on the northern coast of the Straits of Malacca. The consignments are then picked up by local couriers. Alternatively, they are ferried by local fishing boats that frequent southern Thailand to pick up or drop their mostly Thai crews. Equipped with GPS, the boats will place the ganja filled fish-trap, locally called "bubus", at a precise position which is later passed, usually by cellphone, to local couriers. The "bubus" and the ganja consignments are then retrieved for distribution.

Another way of smuggling applicable to both firearms and drugs is the use of boat-to-boat or ship-to-ship transfer. A mother-ship carrying a consignment of drugs or firearms from a port in Myanmar or Thailand, will upon reaching a predetermined location, transfer the consignment to a waiting fishing or a speed boat. The boat will then ferry the consignment to a predetermined port or an isolated landing point. To protect themselves, the syndicates are usually well-armed, including with M16 assault rifles. Most of the successful interceptions have been made after intensive surveillance, intelligence gathering and deception (Salam, 2005).
Among the major interceptions in the south:

- 1990s Penang - authorities intercepted a cache of arms bound for the LTTE in Sri Lanka

- Mid-2003 - Kota Baru, Malaysia, two revolvers intercepted on board an express bus bound for Kuala Lumpur

- End of 2004 - Pulau Panjang, South Thailand. A fishing boat with 50kg of Langkawi-bound cannabis intercepted. During the trafficking operation all telecom lines were jammed.

- End April 2005 - Andaman Sea off Satun beach south Thailand.Thai Navy detained 5 pirates with two M16, three AK47 assault rifles and 2000 rounds of ammunition.

- May 2005 - Andaman Sea, Muang south Thailand.
  Thai authorities detained 5 Indonesians in southern Thai waters suspected of involvement in piracy and gunrunning (Salam, 2005).

- June 2005 - Haadyai, Southern Thailand. Thai Police intercepted Malaysia-bound heroin and methamphetamines hidden in a fertilizer bag at the back of a pick up truck (Zahari, 2005).

- 14 July 2005 - the Malaysian Police intercepted a hijacked coastal tanker, Nepline Delima, that was heading for the Andaman Sea. Why was it heading for the Andaman Sea? (New Straits Times, 2005).

- September 2005 - Port Klang. Malaysian Customs intercepted 7 pistols, a rifle and a shotgun bound for Dubai (New Straits Times, 2005). The interception denotes how easily firearms, believed to originate from Indochina, could be brought into the country and then re-exported.

The arrests and interceptions clearly show the extent to which the Andaman Sea and the peripheral land mass have been used as a thriving drug and arms trading and distribution centre by well organised international syndicates. With their symbiotic relationship and extensive international networks, these syndicates use proceeds from their drug trade to fund gunrunning and human smuggling activities as well as to finance insurgent movements.

As a senior customs officer conceded: "the number of smuggling attempts foiled by the authorities is just the tip of the iceberg. For each arms/narcotics smuggling activity detected, much more simply goes undetected" (Puteh, 2005). If this assessment is to be taken seriously, the impact from these criminal activities is indeed very alarming.

Despite having some of the most stringent drug and firearms laws, narcotics continue to "disable" more than 200,000 Malaysians annually. Some estimate the unofficial figure to be more than double this figure since many wish to avoid disgrace and humiliation to their immediate families, and therefore do not register.

Drug addicts are a liability both to society and to their families. They are not only unproductive, but they also cause problems for many people through their penchant for crimes to support their addiction. Narcotics have been the root cause of most crimes and immoral activities. About 48% of prisoners serving their sentence in Malaysian jails were convicted of drug-related charges.

Most addicts lose their interest in life and neglect their responsibilities, whether to their jobs or their families. There is yet to be a proven effective cure for drug addiction. Statistics have shown that the majority of addicts who have treatment eventually return to their addiction. As such, drug addiction is one of the most serious and complicated social problems facing countries bordering production centres like the "Golden Triangle". It is for this reason that it is regarded as the number one national scourge.

While most of the firearms smuggled from the northern region have been widely used in crimes on land, there is no doubt that some have been used to commit crimes in the Straits of Malacca. In Malaysia, armed robbery and hired killings have been on the rise since the 1990s. From 2000 to June 2001, 413 firearms, consisting mainly of pistols, revolvers and shotguns were seized. Some estimate that the number of firearms that found their way into the country could be as much as 20 times more than those intercepted. The low probability of being caught has mitigated the deterrent value of the death penalty for illegal possession of firearms. Of the 37 sea robbery/piracy cases in the Straits of Malacca in 2004, 22 or 73% were confirmed to involve the use of firearms and most occurred at the northern entrance to the Straits (ICC-IMB, 2005).

Generally, the flow of firearms and drugs occurs in three ways. Firstly, the weapons are smuggled by sea directly from the northern region to the Straits by sea robbers or pirates. These firearms are used to attack high value vessels transiting through the Straits. The villains either target the crew to negotiate for a ransom or are simply happy to get away with their valuables. The more organised attack would go for all - the vessel, their goods and the crew.

Secondly, firearms are smuggled through land routes to Southern Thailand. From there, the consignment is either transferred by land to Malaysia or by sea to Indonesia from where they find their way to "sea robbers" or insurgent groups such as the Gerakan Aceh Merdeka (GAM). Much speculation has been made of rogue GAM members using these firearms to extort for personal gain, or GAM insurgents as a group using the firearms to extort for funds to finance their war.

The decline in sea robbery attacks in the northern part of Sumatra following the end of the GAM insurgency activities seems to reaffirm these speculations. The current insurgency in southern Thailand could result in a similar situation. Drug cartels using drug money could offer arms to insurgents in exchange for assistance in the sale of narcotics or protection of drug traffickers. This nexus is not uncommon in other insurgency-troubled borders. This free-flow of firearms and narcotics in unstable areas close to the Malaysia-Thai border could pose serious security threats to Malaysia and the Straits of Malacca.

Thirdly, the environment of "lawlessness" as a result of rampant drug-trafficking and gunrunning activities perpetuates socio-economic problems onshore. Pressures of socio-economic deprivation subsequently force more to turn to sea for their livelihood. Some would find sea robbery and piracy an attractive option.

Addressing Threats to the Straits of Malacca

National Level
Efforts to tackle transnational criminal activities should first begin with the creation of an effective national criminal enforcement capability. This requires the enhancement of a legal regime and physical law enforcement capacity. While most regional governments have invested substantially in combating drug and firearms related problems in their countries, addressing the menace through reactive means alone, that is, after these lethal contrabands have seeped into their country, is certainly not the best strategy. A forward-looking preventive strategy is needed to tackle the ills at their source, like eradicating the narcotics and firearms production or distribution centres.

Equally important is the need to address the domestic legal lacuna to harmonise and update national laws. "In most littoral states the law is institutionalised at the colonial period of the 19th century" (Singh, 2003). This is not only to ensure that these laws cover the identified threats, but also to ensure that they facilitate international cooperation. Unless there are competent and updated legal procedures that enable the enforcement authorities to put up a strong legal case, and the judiciary to try and pass the appropriate sentences, all the hard work to address the long-term threat from the activities of non-state actors will not be effective.

Then there is a need to strengthen regional law enforcement capabilities. This is more evident in the maritime dimension, possibly due to either a genuine lack of resources or the lack of appreciation of the contribution of the sea in both economic as well as security terms. In developing maritime forces, priority has often been given to navies rather than law enforcement forces. While navies may have most of the traits needed, such as a strong seafaring culture, discipline and an effective command and control structure, they may be over-equipped and too costly to be deployed to carry out routine law enforcement. A professional, dedicated, well-equipped and well-networked maritime law enforcement agency such as the coastguard service, is a more pragmatic option.

Fourthly, efforts must be made to tighten national border checkpoints. This may require the integration in the command of border posts to facilitate the integration and exchange of information among all the agencies involved. For example, while illegal entry is an immigration problem, it could also turn out to be a customs issue if the person concerned is involved in smuggling, or a police problem if the individual is involved in crime.

Regional and International Level

To tackle the transnational criminal activities, any national effort has to be well augmented by trans-boundary cooperation both at international and regional levels. No state on its own, not even a superpower, will be able to effectively act alone. Indeed, for such purposes the United Nations (UN) has instituted numerous international cooperative mechanisms to address the illicit transfer of small arms and narcotics through a UN General Assembly Resolution tabled in December 1999. While the UN-driven initiatives have outlined fundamental action plans, the regional states must collectively 'get their act together'. For ASEAN members there already exists a mechanism for such a purpose. Unfortunately, not all of the Andaman Sea rim states are ASEAN members. Hence, there are gaps in the existing cooperative security mechanism (Bloomfield, 2001).

The most pragmatic multinational security cooperation begins with the sharing of perceptions. This means that what constitutes common trans-boundary threats must be clearly defined. Some may be grouped under various categories like liberation movements, insurgents, belligerents, and so on. Some of these have "now acquired the status of international actors and, thereby, have acquired a legal status under international law". This is especially so for groups whose activities may come under the heading of 'public' motive, as opposed to 'private' motive (Singh, 2003).

Countries may also differ in their definitions of terrorism and insurgency. Although both have been defined in terms of international law, in practical application they are often perceived differently. For example, the Palestinian Liberation Organisation (PLO) struggle to achieve the right to their homeland is viewed by the UN as a legitimate liberation struggle. However, the Israelis, with some Western moral support, view the PLO as a terrorist organisation. In other cases, the issue that often arises is whether insurgents are just groups with legitimate rights. Some of these groups may not be able to articulate their legitimate grievances through formal political institutions and structures to which they are denied access.

Challenges and Prospects

It is acknowledged at the outset that the national and international fight against transnational criminals is full of challenges. At the national level, the main challenge is to develop and sustain law enforcement capability. Even a regional power like India may find it difficult to cope with the task of effectively policing the eastern part of its Indian Ocean. Its Coast Guard still lacks the number of patrol vessels suitable for duty in waters near the coast and in shallow creeks (Asia Times, 2005). Warships, on the other hand, may not be legally, materially or professionally equipped to suppress transnational crimes. And as navies are developed mainly to fight wars at sea, big warships and submarines are of little use against criminals. Getting the regional cooperative undertaking going is yet another major challenge. The web of trans-boundary problems, in some cases worsened by the superpowers geo-politicking, continue to stall genuine cooperation among Andaman Sea rim states.

On the other hand, although regional states may take time to develop their national law enforcement capacities, the outlook seems promising. Indonesia, for example, with a more stable domestic political climate, particularly with the end of GAM insurgency, is able to give more focus and resources to security in the Straits of Malacca. Also, with a more sensitive approach, the littoral states of the Straits of Malacca are also now more amenable to offers of capacity building assistance from the major user states.

Regional powers like India should no doubt play a more active role in the security of the region. But this is not by incessantly making its military presence felt in the Straits of Malacca, as an obsession to display military might would be misconstrued by the littoral states. Its role should rather be focused on ensuring that its part of the Andaman Sea is free of transnational criminal activities. India could act as a regional anchor to crystallise a more effective security cooperation among the regional states. As Dr Vijay Sakhuja, a Senior Fellow at India's Observer Research Group has succinctly said: "The Indian government's approach has been to impress on the littorals that India will not force itself on them but will be ready to provide assets when asked for" (Singh, 2003). India's new Far Eastern

Naval Command (FENC) off Port Blair in the Andaman Islands should be focused on addressing threats of trans-national crimes like sea piracy, small arms proliferation, trans-border migration, smuggling of narcotics, and disaster management, instead of gearing itself to counter the illusionary Chinese threat. Some of the recent regional cooperative trends that offer some promising outlook for regional security cooperation include:

- The recent multinational tsunami relief operations,

- The natural gas exploration/gas pipeline cooperation in the Bay of Bengal,

- The growing security ties between India-Myanmar and other ASEAN countries, and

- The recent Thailand-Laos-Cambodia, agreement on joint suppression of cross-border crimes. Thailand and Laos also agreed to establish joint army and marine patrols (Berlin, 2005).

The resolution of the conflict in Indonesia's Aceh Province provides yet another milestone in reducing threats posed by the illicit transfer of arms within the region. In this case, with no more war in Aceh, there will be less demand for firearms in the Province and hence a reduction in the flow of weapons through the Andaman Sea area. This will eventually reduce the likelihood of these arms falling into the hands of the notorious pirates, sea robbers, or even terrorists.

Conclusion

The northern approaches to the Straits of Malacca encompassing the Andaman Sea and its surrounding rimlands, have for decades been an operating hub for criminal groups like drug traffickers, gunrunners and human smugglers working in symbiosis. The activities of these trans-boundary criminal groups have not only weakened governance in some regional states, but have also endangered their domestic stability. What turned the region into a criminal hub is not purely coincidental. Amongst the contributory factors are: the region's proximity to the drug producing "Golden Triangle"; ex-war surplus arms stockpiles in Cambodia; the lack of effective national governance and inter-state security cooperation. Beyond that, with a well-established marketing network, these illegal trading activities are so lucrative that they are very difficult to stop.

From the interceptions made by the authorities as well as from the analysis of trends of attacks, these criminal activities have no doubt impacted on the security of the Straits of Malacca. Most of the reported brutal attacks involving the use of firearms, happened in the northern part of the Straits.

Threats of transnational criminals from the northern approaches to Straits of Malacca can be addressed in three ways. Firstly, at the national level, regional states should strive to enhance domestic governance and law enforcement. This encompasses the enhancement of national law enforcement capacity and border management through consolidating national laws and institutions as well as equipment. All sources of domestic instability, political or socio-economic, must be identified and addressed through the principle of "justice for all". One of the main challenges to this initiative is inadequate resources. Through proper prioritisation, however, this can be overcome.

Secondly, regional states need to enhance their trans-boundary cooperation. This may include enhancing security cooperation and trans-boundary socio-economic development, especially along the border areas. To ensure effective security cooperation, states may need to identify and agree on what constitutes their common threats, and, at the same time, develop a pragmatic legal and operational mechanism to jointly counter the identified threats. Through close sharing of intelligence, for example, the movement of the criminals can be tracked. Socio-economic cooperation may take the form of joint development and can take the form of regional growth triangles like the Northern Malaysia, Southern Thailand and North Sumatra Growth Triangle. These trans-national initiatives, however, may involve a lengthy political and diplomatic process. And they may require other sensitive trade-offs between the parties involved.

Thirdly, to commit the regional power, in this case India, to secure its part of the criminal infested region as well as to lead and assist the other states in need. It is more meaningful for India to focus on the security of its part of the northern approaches to the Straits of Malacca, rather than projecting its military might in the waterway which is well within the territorial waters of littoral states. Such an act would inevitably be misconstrued by the littoral states. On this issue, it is reassuring to recall the words of Vice Adm (Rtd) Premvir Saran Roy, Co-Chairman of Council for Security Cooperation Asia Pacific (CSCAP), that: ÒIndia will play its role in the Straits of Malacca as desired by the littoral states, not the way the United States wants her to playÓ (Roy, 2006).

While the threats posed by transnational criminal activities in the northern approaches to the Straits of Malacca are indeed serious, it has to be admitted that the proposed solutions are at this stage rather superficial. More research needs to be done, particularly on the politico-diplomatic complexities before a comprehensive and pragmatic national and multinational measure can be formulated.

References

American Foreign Policy Council (2005), *Asia Security Monitor* No. 130. American Foreign Policy Council, 27 June.

Asia Times (2005), 19 October, at http://www.atimes.com

Asia Week (2000), 'Tracking Tigers in Phuket', at http://www.asiaweek.com/asiaweek/magazine/2000/0616/nat.security.html.

Berita Harian (2005), 15 October.

Berlin, D. (2005), 'The Emerging Bay of Bengal', *Asia Times at* http://www.atimes.com

Bloomfield, Lincoln P., Jr. (2001), Assistant Secretary for Political-Military Affairs Monthly briefing on *United Nations Conference on the 'Illicit Trade in Small Arms and Light Weapons in All its Aspects', 28 June.*

Capie, D. (2002), 'Small Arms Production Transfers in South East Asia', Canberra Papers *on Strategy and Defence*, No. 146.

CIA web site.

CSIS (1998), Research Report on 'Combating Illicit Light Weapons Trafficking: Developments and Opportunities', *British American Security Information Council*, London, January.

Das, N. R. (2004), 'Narco-Politics and Islamic Society: A Case Study of Pakistan', 8 January, available at http://www.defenceindia.com/defenceind /research.html.

Dawn (2005), 'India/Burma: Naga Tribes Demand Homeland on India, Myanmar border", 31 August.

FTU-B (2003), 'Migrant Workers from Burma', *Federation of Trade Unions-Burma: Responsibilities and Rights*, 6 January.

Gurung, Madhu and Ramtanu Maitra (1995), 'Insurgent Groups in Northeast India', *Executive Intelligence Review*, 13 October.

Mahapatra, A. B. (2005), 'Andaman Faces Kargil-type Invasion', *The Public Affairs Magazine*, News Insight, 5 April.

National Drug Agency statistics data.

New Straits Times (2005), 15 July.

Pongsudhirak, Thitinan (2005), *New Straits Times*, 22 September.

Puteh, O. (2005), Interview with Mr Omar bin Puteh, Assistant Director of Customs, Kedah, 31 May.

Raman, B. (2002), 'Control of Transnational Crime and War against Straits Times Terrorism', *Indian Defence Review*, Vol. 17 (2), April-June.

Roy, P. S. (2006), VAdm (Rtd) Premvir Saran Roy, Co-Chairman, *4th CSCAP Meeting*, Renaissance Hotel, Kuala Lumpur, 27-28 May.

Rumley, D. (1991), 'Society, state and peripherality: the case of the Thai-Malaysian border landscape', chapter 7 in D. Rumley and J. V. Minghi, eds., *The Geography of Border Landscapes*, London : Routledge, pp. 129-151.

Salam (2005), Interview with DSP Abdul Salam bin Abdul Halim, Deputy Commander, Malaysian Marine Police, Langkawi, 2 June.

Samanta, Pranab Dhal (2005),, *The Indian Daily Express, 1 July*.

Singh, K. R. (2003), 'Maritime Violence and Non-state Actors: with special reference to the Andaman Sea and its Environment', *Dialogue*, Vol. 4 (4), April-June.

Solahudin, H. (2005), Interview with Mr Haji Solahudin, Director of Customs, Kelantan, 7 June.

The Burma Net News (1995), 26 April, Issue #157.

The Hindustan Times (1998), 12 September.

Travelblog (2005), at http://www.travelblog.org/world/th.int.html

UN Report on Trafficking scene in South and South West Asia, Chapter 4 available at http://www.unodc.org/pdf /india/ccch4.pdf

Utusan Malaysia (2005), 5 May.

Chapter 15

**Indian Ocean Sea Lane Security and Freedom of Navigation:
Legal Regimes and Geopolitical Imperatives**

Manoj Gupta

Introduction

The history of the law of the sea is a narration of the struggles for and against the doctrine of the freedom of navigation. Historical research clearly demonstrates that freedom has no static content 'a priori', but is subject to continuous, at times even violent, changes. This chapter will explore how new legal regimes and geopolitical imperatives can provide for freedom of navigation and sea lane security as points on the same continuum in the relationships among flag states and coastal states in the Indian Ocean.

Geography

The Indian Ocean extends over one fifth of the world's total ocean surface. Its strategic waterways account for the transportation of the highest tonnage of goods globally Ð half the world's crude oil, container shipments and a third of the bulk cargo. The Indian Ocean Region comprises 30 littoral states, 5 island states and 13 other island territories, and 13 landlocked countries. According to the UN classification, countries which have access to the Indian Ocean only via a coastal state bordering the Indian Ocean and whose security is affected by development in those states form the Indian Ocean Region (Braun, 1983). By this definition, including the five Central Asian States of Kyrgyzstan, Tajikistan, Uzbekistan, Kazakhstan and Turkmenistan amongst the landlocked states - whose maritime access lies either to the Baltic Sea or the Black Sea through the Caspian Sea or to the Indian Ocean, through Iran, Afghanistan and Pakistan - there are 53 member countries of the UN and 13 other island territories in the Indian Ocean Region (Appendix). In the maritime context, such a broad definition of an Indian Ocean Region is the favoured geopolitical definition (Bouchard, 2004). The 53 countries represent over one-third of the world's population; they account for two-thirds of the world's proven oil reserves, a third of its natural gas, 90% of its diamonds, 60% of its uranium and 40% of its gold. It is not surprising, therefore, for Fernand Braudel to have observed that the Indian Ocean was a *"Weltwirtschaft"* or 'World Economy' with India at the centre. In the context of 'Sea Lane Security in the Indian Ocean' who is included in a regional definition and who gets defined out will depend on changing economic, cultural, and political relationships and linkages (Bergin, 1996).

Law of the Sea

The law of the sea emerges as one of the oldest components of international law (Brown, 1994). It was the third UN Convention on the Law of the Sea (UNCLOS-III) in 1982 that consolidated all past treaties, codified customary law and put in place new law for new issues. In one stroke, the law of the sea evolved into hard law from soft law. Through the 1990s, the legal regime for oceans has continued to be strengthened through a number of conventions, agreements and programs of action that were negotiated, adopted or entered into force to address different issues. As many as 40 international agreements and programs have been negotiated, adopted or entered into force in an effort to promote peace and security, as well as the sustainable development of the oceans. The range of international agreements dealing with all aspects of maritime affairs have contributed to international norm setting for ocean management, but implementation at the regional and national level continues to be a challenge.

Maritime Security

From a maritime security perspective, at the end of the Cold War the maritime domain has witnessed the emergence of new categories of security challenges Ð environmental degradation, resource scarcity, transnational crime, piracy, drug trafficking, illegal immigration, terrorism and so on. Many of the issues are beyond the capacity of individual states to tackle alone even though tough, independent measures can be taken to reduce their vulnerability to certain threats. The empowerment of transnational non-state actors necessitates international cooperation to deny them access to sanctuaries, weaponry, finances and other resources. Consequently, despite the growth in sea denial capability of a number of states there is increasing interdependence for maritime security.

Geopolitics

From a geopolitical perspective, effective forms of ocean management to address issues such as SLOC security have been rendered more complex by the increased number of nation-states and their differing interpretations of the UN Laws of the Sea. The major maritime powers benefit from the current arrangements or from the absence of them. This is even more so when traditional negotiating blocks Ð like for disarmament, development, trade and investment Ð are much less in evidence in the case of the oceans. Further, the globalised economic ideology of states is placing heavy reliance on the market and the role of private enterprise and therefore states are not receptive to regulatory mechanisms for the oceans. The absence of regulatory mechanisms has led to the abuse of oceans. Growing awareness of the problems of the oceans has not yet permeated to individual behaviour. States left to their own devices are unable to dispense answers to questions relating to peace, security, equity and the environment.

Freedom of Navigation on the High Seas

Comprising the world's largest expanse of common spaces, the high seas traditionally nearly touched the shores of the world's coastlines. Until 1958, the high seas were viewed, as all parts of the sea not included in the territorial sea or the internal waters of a state (Brown, 1994). In 1782, the Italian jurist, Galiani, suggested that, for the purposes of neutrality, the breadth of territorial waters should be a maritime belt coterminous with a 'cannon shot' (Anand, 1982). It was in 1793, during the war between England and France, that US Secretary of State, Jefferson, equated the range of a cannon ball to one sea league or 3nm (Anand, 1982). England, in its Territorial Waters Jurisdiction Act of 1878 recognised this as a result of three decisions of Lord Stowell between 1800 and 1805 (O'Connell, 1982). Ever since, various states have submitted proposals that have varied from 3 to 200nm (Vogler, 1995).

UNCLOS III marked an historically important shift in the balance between 'control and regulation' and the 'freedom of navigation on the high seas'. The high seas, under article 86 of UNCLOS III, came to be defined as 'all parts of the sea that are not included in the exclusive economic zone, in the territorial sea or in the internal waters of a state, or in the archipelagic waters of an archipelagic state'. In other words, one-third of the world's ocean space was now placed under the national jurisdiction of the coastal states for exploring, exploiting, conserving, and managing the living and non-living resources, such as fisheries and hydrocarbons (Forbes, 1995). Thus, the freedoms making up the 'freedom of navigation on the high seas' established as a fundamental principle of international law (Churchill and Lowe, 1992), had been subjected to varying control and regulation by specifying the rights and responsibilities of all states in the different maritime zones of jurisdiction. For that matter, it is not difficult to find reasons for coastal states to even challenge the notion that the "residual" high seas regime should go largely unregulated (Bergin, 1995).

Legal Regimes

The international community's effort to better regulate the oceans over environmental concerns, and since the terrorist attacks on the World Trade Centre in September 2001, security concerns have resulted in new post-Cold War legal regimes. Environmentally, the grounding of *Exxon Valdez* on the night of March

24, 1989 initiated a debate on the navigational freedom of vessels carrying dangerous cargo and on a coastal state's interest in protecting its marine environment. It triggered a national and international response to avoid recurrence at the cost of freedom of navigation. The US enacted the Oil Pollution Act 1990 which set into motion a process for national response plans, provisions for fixing liability, determination of removal and damage costs and the phasing out of single hulled tankers (US EPA web site). The International Convention on Oil Pollution Preparedness and Response adopted in November 1990 called for establishing measures for dealing with pollution incidents, either nationally or in cooperation with other countries (IMO web site). The 1992 amendment to the International *convention* for the Prevention of Pollution from Ships (MARPOL 73/78) made it mandatory to have "double hull" tankers (IMO web site). The sinking of the Erika off the coast of France in December 1999 led to a new accelerated phase-out schedule for single-hull tankers (IMO web site). The break-up of the oil tanker, *Prestige*, off the Spanish coast in November 2002 got Spain and France to enact a law requiring tankers passing through their EEZs to give prior notification of their cargo, destination, flag and operators, with single hulled tankers more than 15 years old determined to be not seaworthy being expelled from the EEZs (Van Dyke, 2005, pp. 109). Interestingly, in 1998, India had issued a similar notice for all vessels carrying dangerous cargo and entering the Indian EEZ to give prior notification, which was protested to by France, the US and some other countries.

From a security perspective, countries have moved swiftly to address proliferation, terrorism and maritime security, including piracy, drug trafficking and arms trafficking. On 31 May 2003, the US initiative supported by select countries outlined the Proliferation Security Initiative (PSI) to prevent the flow of WMD, their delivery systems, and related materials on the ground, in the air, and at sea, to and from states and non-state actors of proliferation concern (US State Department web site). The PSI interdiction principle sought to expand international law to permit active military intercepts changing the law permitting freedom of navigation on the high seas (Van Dyke, 2005, pp. 119). After 9/11, a comprehensive maritime security regime was quickly introduced with a number of amendments to the 1974 Safety of Life at Sea convention (SOLAS). One of most important consequences is the International Ship and Port Facility Security (ISPS) code brought into force on 1 July 2004 (IMO web site). The 1988 Suppression of Unlawful Acts (SUA) convention and its protocol are aimed at taking appropriate action against persons committing unlawful acts against ships and offshore fixed platforms. A diplomatic conference was held in October 2005 to adopt amendments aimed at strengthening the SUA treaties in order to provide an appropriate response to the increasing risks posed to maritime navigation by international terrorism (IMO web site).

Geopolitical Imperatives

The geopolitical importance of the Indian Ocean as a region has at times been underestimated (Rumley and Chaturvedi, 2004). From a global perspective, the Indian Ocean does not comprise a "true security system" (ibid). However, it can be said that the geopolitical imperatives involving the right to exploit fisheries and hydrocarbon resources in the Exclusive Economic Zone, environmental protection and maritime security issues have by compressing the breadth of the High Seas affected freedom of navigation in the Sea Lanes and created a security system for shipping within the Indian Ocean.

As a consequence, maritime boundary disputes and overlapping jurisdictional claims appear to threaten the closure of ocean spaces traditionally used to ply free trade. The maritime claims that could impact on sea lane freedom of navigation include (Tsamenyi and Mfodwo, 2001):

(a) Unrecognised claims to historic waters.
(b) Claims to jurisdiction over maritime areas in excess of 12nm.
(c) Contiguous zones at variance with article 33 of UNCLOS.
(d) EEZ claims not consistent with Part V of UNCLOS.
(e) Continental shelf claims inconsistent with Part IV of UNCLOS.
(f) Territorial sea claims overlapping straits used for international navigation.
(g) Archipelagic claims which are not in conformity with UNCLOS.

Freedom of Navigation and Sea Lane Security

Over the last decade there have been substantial pieces of work on SLOC security that examine:

(a) SLOCs as economic lifelines.
(b) UNCLOS and the freedom of navigation principle.
(c) SLOC insecurities and the centrality of choke points.
(d) Piracy and, more recently, maritime terrorism.
(e) Wide-ranging recommendations for regional SLOC security cooperation.

In understanding the geostrategic implications of the current form of maritime transportation, Jean-Paul Rodrigue provides four perspectives - conquest, cooperation, competition and control (2004). Transport technology was initially a means of conquest to acquire and conquer oceans, territories and resources (ibid). Today, the military technology spectrum has separated from commercial technology. The military has little interest in committing high value warships to low end roles of commerce protection and led to the mushrooming of coast guards. Simultaneously, as a result of liberalised commercial operations, shipping and ports want nothing to do with the military in protection and are willing to have their own security arrangements. Secondly, technological advances in shipbuilding have led to specialised and extremely large merchant ships catered to meet specific demands of the international supply chain. As a result, the military is happy to develop its own sealift capability. What these changes in transport technology have done is that there is less synergy between a coastal state's naval fleet and the merchant fleet.

The shift from traditional national transport companies to the new form of international transport competition for reducing costs has introduced Flags of Convenience (Ibid) and substandard ships. Coupled with the lack of military interest in merchant ships it has led to the virtual demise of most national merchant fleets. This has increased SLOC insecurities and has had a direct impact on a coastal stateÕs ability to sustain sea lane security and security of its coastlines within the overriding desire to preserve the freedom of navigation regime.

The international nature of trade has favoured agreements over different aspects involving access to infrastructures or setting standards leading to the emergence of economic blocs such as the EU, NAFTA and APEC (Ibid). The focus on cooperation to gain economic advantage and retain the present international supply chain makes it inescapable to have regional cooperation for SLOC security. However, regional SLOC security cooperation efforts are yet to materialise in the same way as economic blocs have emerged because countries are willing to sidestep political alignment on security constraints in their effort to promote and preserve economic interests.

The control of strategic places is seen as being vital to international transportation of trade (Rodrigue, 2004). Three of the six most important straits - Babel Mandeb, Dover, Dardanelles, Gibraltar, Hormuz and Malacca - on which world shipping concentrates, link the Indian Ocean to the other oceans. The straits of Babel Mandeb and Hormuz are major oil transit choke points with no alternative routes. The only viable alternative via Lombok to the Malacca Strait adds nearly 1200nm to a voyage from the Persian Gulf to Japan. The principal reason for Hugo Grotius *"mare liberum"* or the concept of "Freedom of the High Seas" was to remove and prevent restrictions on international trade. Ever since, major maritime powers for their own security interests have restated the concept to imply as much 'free and open' military use of the oceans as possible. In the case of choke points, the imperative to preserve the freedom of navigation regime to serve military interests is in conflict with the bordering coastal states' responsibility to regulate and exercise greater control in its effort to address SLOC insecurities.

What this means is that national polices concerning the geostrategic aspects of conquest, competition, cooperation and control portend that:

(a) Merchant fleets get much less attention for military protection and are increasingly proving to be of little military use for sealift requirements.
(b) States are giving less importance to national shipping. As the global phenomenon of shipping gains momentum, national merchant fleets are on the decline and could soon vanish for some states.
(c) Cooperation between states for ship security and as a consequence for sea lane security is difficult to achieve because such cooperation is politically viewed through the lens of economic benefit.
(d) Freedom of navigation is a military necessity and international transportation of trade the pretext.

It is within this geostrategic realm that the legal regimes and geopolitical imperatives have evolved. As a consequence, freedom of navigation has been generally affected in three ways:

(a) Geographic closure of ocean spaces.
(b) Enforcement mechanisms that control and regulate shipping in an effort to address maritime security and protect the marine environment.
(c) Security measures that constrain shipping in an effort to address proliferation and terrorism.

Contrary to what is normally perceived, there is hardly any risk of the geographic closure of ocean spaces in the Indian Ocean and it is unlikely that maritime boundary disputes will lead to conflict thereby impinging on freedom of navigation. Governments in the Indian Ocean have demonstrated the will and determination to put aside past differences and negotiate on the concerns and issues that relate to their adjacent seas (Forbes, 2004). This is evident from the recent happenings over the critical India-Pakistan Sir Creek dispute. The two countries commenced a joint survey on 24 January 2005 (Deccan Herald, 2005) and on 28 May 2005 commenced talks to review the joint survey with India proposing a seaward approach to demarcate the maritime boundary between the two countries pending formalisation of the boundary in Sir Creek (The Hindu, 2005). It is stated that, in the Indian Ocean, it is only the enforcement mechanisms and security measures imposed by states for sea lane security that could impinge on the freedom of navigation, for instance to affect:

(a) Innocent passage through the territorial waters and contiguous zones of Sri Lanka and Maldives.
(b) Transit passage through the International straits of Hormuz, Aden and Malacca.
(c) Archipelagic Sea Lane passage through Indonesia.

(d) Passage through the EEZ of India.

(e) Navigation on the High Seas due to actions of the US and other select maritime powers.

What this means is that the legal regimes to address the environmental and security concerns circumscribe the Freedom of Navigation principle. Furthermore, the geopolitical imperatives of maritime boundaries in the coastal states' effort to manage and exploit resources and protect the marine environment dictate mechanisms for surveillance, monitoring and control, including boarding, inspection, arrest and judicial proceedings that, although beneficial to sea lane security, impinge on freedom of navigation. The emerging legal regimes and geopolitical imperatives reopen the debate between *Mare Liberum* and *Mare Clausum*.

As stated earlier, freedom of navigation was practiced by states to promote free trade but subsequently came to be argued by maritime powers to transit choke points and traverse across the High Seas as a matter of right than at the sufferance of the coastal and island states (Ibid). What has not been stated is that freedom of navigation passed into international customary law because of the power enjoyed by major maritime powers to be able to enforce it. Should coastal states develop the power to enforce, state practice could pass into customary international law, should the majority of the 149 coastal states from the 191 member countries of the UN adopt similar state practice. Rights of passage have essentially come into existence as a rule of international customary law (Ibid). The US Freedom of Navigation Program commenced in 1979 is but one example to counter state practice and provides expression and definition to the generalities and ambiguities of UNCLOS that could otherwise pass into international customary law.

UNCLOS had actually attempted to balance traditional navigational freedoms and coastal states' rights to control and monitor the movement of hazardous cargo (Wecker and Wesson, 1993) and fishing vessels, including boarding, inspection, arrest and judicial proceedings, in the exercise of coastal states' sovereign rights to protect the marine environment and to explore, exploit, conserve and manage the living resources in the exclusive economic zone (Van Dyke, 2005, pp. 108).

The real issue here is whether freedom of navigation and sea lane security remain on the same continuum. To take Geoffrey Till's review of maritime power in the 20th century (Till, 1998) forward, one can argue that, in the new millennium, the political, economic, legal environmental and technological constraints necessitate a shift from the concepts and attitudes of maritime powers as advocated by Mahan and Corbett to those concerned much more with collaboration, coordination and cooperation. According to one Indian analyst, there is a near total absence of a multilateral approach to safeguard sea lanes in the Indian Ocean (Sakhuja, 2004). Since 2001, SLOC security has witnessed a sea change in the attitudes and thinking of navies and coast guards. The beginnings of cooperative approaches are visible, but sea lane security requires a shift from governmental unilateral and bilateral approaches to system-wide arrangements that look at the supply chain in totality (Bergin and Bateman, 2005). This requires active participation of the logistics and transport industry and a clear mandate to a single agency in government to secure the entire supply chain. Some of the salient issues involved in securing the supply chain include:

(a) Port State Control (PSC) gridlock.
(b) Monitoring and control system.
(c) Surveillance and information exchange
(d) Agreed procedures for boarding, inspection, arrest and judicial proceedings.

The guidelines for coastal states in the Indian Ocean to implement these system wide arrangements include:

(a) Collaboration among coastal states to set up a PSC gridlock up to 24nm or beyond. The ISPS code is a step in this direction. However, it further needs agreed procedures for boarding, inspection, arrest and judicial proceedings to be spelt out. For instance, the procedure to be adopted in dealing with Flags of Convenience could be different from the procedure for flag states. A Flag of Convenience report should include a mandatory clause for declaring the nationality of ownership. Coastal states also need to collaborate on other aspects such as crew verification procedures and ISPS implementation assessment programs. Another critical area for collaboration is developing national databases of vessels > 300 GRT and ensuring that voyages by these vessels involving more than one PSC are handed over from PSC to PSC.

(b) Coordination among coastal states beyond the PSC grid lock up to the outer limit of the EEZ for monitoring and control. For example, coastal states would need to coordinate not just joint patrols, exercises and training, but also coordinate in developing contingency plans to deal with different threats to SLOC security.

(c) Cooperation between coastal and flag states beyond the EEZs of coastal states for surveillance and information exchange. For example, coastal states need to cooperate on networking adjacent Search and Rescue Regions (SRRs) in the Indian Ocean and devising mechanisms of mandatory reporting by merchant ships either using the Automatic Identification System or other communicative methods.

Sam Bateman has comprehensively detailed the need to develop similar mechanisms and to implement them for confidence-building measures in overcoming suspicions concerning intentions of neighbours, interoperability aspects and technological levels or limitations of navies (Bateman, 2004). Anthony Bergin has examined East Asian naval developments and has concluded that countries need to tread warily with naval cooperation, lest it stimulate controversy and tension, while gradually moving to exploit opportunities for building confidence at sea (Bergin, 2002). Therefore, the question really is whether or not coastal and flag states can change their mindsets and put aside their suspicions of each other's intentions. Theoretically, and possibly practically too, the system-wide arrangements for sea lane security are possible only when this change occurs in the mindsets of state parties involved. It means acceptance of the need to collaborate, coordinate, and cooperate in introducing mechanisms for PSC, surveillance, monitoring, control, and information exchange and evolving procedures for boarding, inspection, arrest, and judicial proceedings. Secondly, any political alignment between states on sea lane security would need to recognise that freedom of navigation will rest within the system-wide arrangements outlined.

Conclusion

The demand on sea lanes is only increasing. The Indian Ocean is the world's busiest highway. Since medieval times, sea lane security has concerned states in different ways but always from piratical attacks and this will continue to be a perennial problem. Until the 19th century, commerce raiding was an economic attitude which measured economic success in terms of the accumulation of assets by the state. In the early 20th century, commerce protection was a military response to the unlimited German submarine campaign. However, in the 1970s and 1980s, the Cold War was a Western politico-strategic response to the Soviet threat. In the 21st century, threats to sea lane security stem less from direct state action against ships than from indirect action to prevent environmental damage and maritime terrorism, and the potential of regional instability as a consequence of new disputes over maritime boundaries.

UNCLOS is a response to the rapid pace of technological, social, environmental, economic and, more importantly, political changes that have altered the equation of relationships at sea. Technological advances brought in the need to regulate and control ships in high-density traffic areas. Socially, the world populace started moving towards the coastline. Unthinkable environmental damage turned to reality with disasters like the Torrey Canyon. Economically, controlling the mineral wealth of the oceans became a necessity for the sustenance of the worldÕs economies. Politically, a host of states, particularly coastal states, had gained their independence (Brown, 1994).

UNCLOS III has provided for the change in emphasis from 'freedom of navigation on the high seas' to 'control and regulation'. Coastal states have been quick to exercise their rights by establishing their maritime zones for control and regulation. The effectiveness of control and regulation, however, depends not upon establishing jurisdiction in the coastal state legislation, but in actually exercising such jurisdiction in practice (Vidas and Ostreng, 1999). SLOC security, debated for so long, continues to be a challenge and it raises a larger question as to whether the management of the oceans in common is any longer effective and sustainable (Bergin, 1995, p. 198; Appendix). Thus, there is a need to recognise the associated responsibilities if the rationale of UNCLOS III changing the emphasis from 'freedom of navigation on the high seas' to 'control and regulation' is to be not only understood but also to be met. Similarly, there is a need to recognise that along with the 'freedom of navigation on the high seas', that there is a corresponding responsibility for those seeking to exercise that right. Due to the ever-increasing demands and abuses of the seas, the international community has been obliged to ensure the effective exercise of freedom of navigation by controlling and regulating it even more closely. In the Indian Ocean this means that sea lane security is not only possible, but freedom of navigation can be exercised within the ambit of the system-wide arrangements for the purpose.

References

Anand, R. P. (1982), *Origin and Development of the Law of the Sea*, 1st edition, Netherlands: Martinus Nijhoff Publishers.

Bateman, S. (2004), 'Freedom of Navigation and Indian Ocean Security: A geopolitical analysis', in *Geopolitical Orientations, Regionalism and Security in the Indian Ocean*, edited by Dennis Rumley and Sanjay Chaturvedi, South Asian Publishers, New Delhi.

Bergin, A. (1995), 'The High Seas Regime - Pacific trends and developments, in *The Law of the Sea in the Asia Pacific Region*, edited by James Crawford and Donald R. Rothwell, Martinus Nijhoff Publishers, Dordrecht, The Netherlands.

Bergin, A. (1996), 'Defining the Asia Pacific Region, in *The Role of Security and Economic Cooperation Structures in the Asia Pacific Region*: Indonesian and Australian Views, edited by Hadi Soesastro and Anthony Bergin, Centre for Strategic and International Studies, Jakarta, Indonesia.

Bergin, A. (2002), 'East Asian naval developments - sailing into rough seas', *Marine Policy*, Vol. 26.

Bergin, A. and Bateman, S. (2005), *Future Unknown: The Terrorist Threat to Australian Maritime Security*. The Australian Strategic Policy Institute, Canberra, Australia.

Bouchard, C. (2004), 'Emergence of a New Geopolitical Era in the Indian Ocean: Characteristics, Issues and Limitations of the Indianoceanic Order', *in Geopolitcal Orientations, Regionalism and Security in the Indian Ocean* edited by Dennis Rumley and Sanjay Chaturvedi, South Asian Publishers, New Delhi.

Braun, D. (1983), *The Indian Ocean Region of Conflict or Zone of Peace*, 1st edition, Oxford University Press: New Delhi.

Brown, E. D. (1994), *The International Law of the Sea*, Two Volumes 1st edition, Aldershot: Dartmouth Publishing.

Churchill, R. R. and Lowe, A. V. (1992), *The Law of the Sea*. 2nd edition, Manchester: Manchester Univ. Press.

Deccan Herald (2005), 'India and Pakistan begin joint survey of Sir Creek', 24 January.

Forbes, V. L. (1995), *The Maritime Boundaries of the Indian Ocean Region*. 1st edition, Singapore: Singapore Univ. Press.

Forbes, V. L. (2004), 'Indian Ocean Maritime Boundaries: Jurisdictional Dimensions and Cooperative Measures', in *Geopolitical Orientations, Regionalism and Security in the Indian Ocean*, edited by Dennis Rumley and Sanjay Chaturvedi, South Asian Publishers, New Delhi.

Grunawalt, R. J. (2000), 'Freedom of navigation in the post-Cold War era', in *Navigation Rights and Freedoms: the New Law of the Sea*, edited by Donald R. Rothwell and Sam Bateman, Kluwer International, Netherlands.

IMO web site: http://www.imo.org/home.asp

OÕConnell, D. P. (1982), *The International Law of* the Sea Vol. I. 1st edition, London: Oxford Univ. Press.

Rodrigue, J.-P. (2004), *Transport Geography on* the Web, Hofstra University, Department of Economics & Geography, http://people.hofstra.edu/geotrans

Rumley, D. and Chaturvedi, S. (2004), 'Changing geopolitical orientations, regional cooperation and security concerns in the Indian Ocean', in *Geopolitical Orientations, Regionalism and Security in the Indian Ocean*, edited by Dennis Rumley and Sanjay Chaturvedi, South Asian Publishers, New Delhi.

Sakhuja, V. (2004), 'Contemporary Piracy, Terrorism and Disorder at Sea: Challenges for Sea Lane Security in the Indian Ocean', in *Geopolitical Orientations, Regionalism and Security in the Indian Ocean*, edited by Dennis Rumley and Sanjay Chaturvedi, South Asian Publishers, New Delhi.

The Hindu (2005), 'Dialogue on Sir Creek begins', 29 May.

Till, G. (1998), A review: 'Maritime power in the 20th century' in the *Maritime Power in the 20th Century: The Australian experience* edited by David Stevens, Allen & Unwin, NSW, Australia.

Tsamenyi, M. and Mfodwo, K. (2001), 'Analysis of contemporary and emerging navigational issues in the Law of the Sea', Seapower Centre and Centre for Maritime Policy, *Working Paper*, No 8, RAN Seapower Centre, Australia.

US EPA web site: http://www.epa.gov/oilspill/opaover.htm

US State Department web site: http://www.state.gov/t/np/rls/fs/46839.htm.

Van Dyke, J. M. (2005), 'The disappearing right to navigational freedom in the exclusive economic zone', *Marine Policy*, Vol. 29.

Vidas, D. and Ostreng, W. (1999), *Order for the Oceans at the turn of the Century.* 1st edition, The Hague: Kluver Law International.

Vogler, J. (1995), *The Global Commons.* 1st edition, New York: John Wiley and Sons.

Wecker, M. and Wesson, D. M. (1993), 'Seaborne movements of Hazardous Materials' in Van Dyke, J M., Zaelke, D. & Hewison, G., eds, *Freedom for the Seas in the 21st Century: Ocean Governance and Environmental Harmony.* 1st edition, Island Press, Washington DC.

APPENDIX

Indian Ocean States

SNo	Country	Date of UN Membership
	South East Asia and Malacca Straits	
1	Australia	01 Nov 1945
2	Indonesia	28 Sep 1950
3	Myanmar	19 Apr 1948
4	Malaysia	17 Sep 1957
5	Singapore	21 Sep 1965
6	Thailand	16 Dec 1946
	South Asia	
7	Afghanistan	19 Nov 1946
8	Bangladesh	17 Sep 1974
9	Bhutan	21 Sep 1971
10	India	30 Oct 1945
11	Nepal	14 Dec 1955
12	Pakistan	30 Sep 1947
13	Sri Lanka	14 Dec 1955
	Persian Gulf	
14	Bahrain	21 Sep 1971
15	Iran	24 Oct 1945
16	Iraq	21 Dec 1945
17	Kuwait	14 May 1963
18	Oman	07 Oct 1971
19	Qatar	21 Sep 1971
20	Saudi Arabia	24 Oct 1945
21	United Arab Emirates	09 Dec 1971
	Horn of Africa and Red Sea	
22	Djibouti	20 Sep 1977
23	Egypt	24 Oct 1945
24	Ethopia	13 Nov 1945
25	Eritrea	28 May 1993
26	Israel	11 May 1949
27	Jordan	14 Dec 1955
28	Somalia	20 Sep 1960
29	Sudan	12 Nov 1956
30	Yemen	30 Sep 1947
	Eastern and Southern Africa	
31	Botswana	17 Oct 1966
32	Burundi	18 Sep 1962
33	Kenya	16 Dec 1963
34	Lesotho	17 Oct 1966
35	Malawi	01 Dec 1964
36	Mozambique	16 Sep 1975
37	Rwanda	18 Sep 1962
38	South Africa	07 Nov 1945
39	Swaziland	24 Sep 1968
40	Tanzania	14 Dec 1961
41	Uganda	25 Oct 1962
42	Zambia	01 Dec 1964
43	Zimbabwe	25 Aug 1980
	Island States	
44	Comoros	12 Nov 1975
45	Madagascar	20 Sep 1960
46	Maldives	21 Sep 1965

Chapter 16

Joint Cooperation Zones to Enhance Maritime Security

Robert Beckman

Introduction

Everyone agrees that international cooperation is necessary to enhance maritime security on the major routes used for international navigation between the Indian Ocean and East Asia. However, cooperation on maritime security is difficult because it often conflicts with one of the two fundamental principles of international law governing the jurisdiction over ships in maritime areas.

The first principle is that ships on the high seas are subject to the exclusive jurisdiction of the flag state, save in certain exceptional cases expressly recognised in international treaties. This principle is set out in Article 92 of the 1982 United Nations Convention on the Law of the Sea (1982 UNCLOS). Article 58(2) of 1982 UNCLOS provides that this principle also applies to foreign ships in the exclusive economic zone (EEZ). This is because the exclusive economic zone is not under the sovereignty of the coastal state, but is a special resource zone in which the coastal state has sovereign rights to explore and exploit natural resources (Article 55-56, 1982 UNCLOS).

The second principle governing jurisdiction over foreign ships is a general principle of international law relating to areas under the territorial sovereignty of states. The principle is that no state may exercise police power in an area under the territorial sovereignty of another state without the consent of that state. This principle would be applicable in the territorial sea and in archipelagic waters, which are the two maritime zones under the territorial sovereignty of the coastal state. No state may exercise patrol or exercise police power over ships in these maritime zones without the express consent of the coastal state.

In this chapter, it will be argued that it is possible to establish joint cooperation zones involving foreign patrol vessels in order to enhance the security of sea lanes of communication and combat illegal activities. First, I will examine the establishment of joint cooperation zones in areas outside the sovereignty of the coastal state such as in exclusive economic zones. I will argue that it is relatively simple to establish joint cooperation zones in such areas. Second, I will examine the establishment of joint cooperation zones in areas subject to the sovereignty of coastal states such as the territorial sea. I will argue that it is possible to establish a joint cooperation zone in such areas without undermining the sovereignty of the littoral states if the joint cooperative zone arrangement is established in accordance with certain principles. Third, I will examine the question of whether it would be possible to establish a joint cooperation zone in the Straits of Malacca and Singapore. I will argue that it is possible to establish a joint cooperation zone in the Straits involving the joint patrols by foreign patrol vessels in a manner that does not undermine the sovereignty of the littoral states.

The phrase 'joint cooperation zone' should not be confused with 'joint development zone', the latter being zones in which two or more states agree to cooperate by jointly developing the resources in a marine area. Joint development zones are usually created by coastal states in areas of overlapping claims when they are unable to reach agreement on a maritime boundary. In this chapter, the phrase 'joint cooperation zone' refers to a defined ocean area in which two or more states agree to cooperate for the purposes of controlling criminal activities at sea which threaten maritime security such as piracy, armed robbery against ships, maritime terrorism, people smuggling, drug smuggling or arms smuggling.

Joint Cooperation Zones in Areas Outside the Sovereignty of the Coastal State

The high seas are open to all states and are governed by the principle of freedom of the seas (Article 87, 1982 UNCLOS). No state may purport to subject any part of the high seas to its sovereignty (Article 89, 1982 UNCLOS). The exclusive economic zone is a specific legal regime adjacent to the territorial sea in which the rights and jurisdiction of the coastal state, and the rights and freedoms of other states, are as set out in 1982 UNCLOS (Article 55). The rights, jurisdiction and duties of the coastal state are set out in Article 56, and the rights and duties of other states are set out in Article 58. Article 58(2) also provides that the principles of jurisdiction governing ships on the high seas apply in the exclusive economic zone, including the provision providing for exclusive jurisdiction of the flag state and the provisions governing piracy.

States can establish maritime zones of cooperation to enhance maritime security in ocean spaces that are within the high seas and exclusive economic zone. These areas are sometimes referred to as "areas seaward of the outer limits of the territorial sea", meaning that they are areas outside the territorial sovereignty of any state. In such areas cooperating states could undertake patrols or other activities to enhance maritime security. This can include the boarding and search of vessels suspected of engaging in criminal activities at sea. However, three limitations would apply.

First, no boarding and search of a suspect vessel should be undertaken without the consent of the flag state. Second, the illegal activities subject to the joint patrols could not include illegal fishing or exploitation of the resources of the exclusive economic zone because the coastal state has exclusive jurisdiction over such activities. Third, the states participating in the joint cooperation zone would have to have due regard for the rights and duties of coastal states in the exclusive economic zone (Article 58(3), 1982 UNCLOS). This means that ships participating in the cooperative arrangements could not unreasonably interfere with fishing or other economic activities of the coastal state in the exclusive economic zone.

The Proliferation Security Initiative (PSI) is a response led by the United States to the growing challenge posed by the proliferation of weapons of mass destruction (WMD). It envisages that cooperating states will work together to intercept ships suspected of carrying WMD. It will be possible for states participating in the PSI to cooperate to intercept ships suspected of carrying WMD or their component parts in areas seaward of the outer limit of the territorial sea. However, any boarding and search of a foreign vessel on the high seas or in the exclusive economic zone under the PSI would be subject to the principle of flag state consent. Consequently, the United States has made arrangements for expedited consent in its agreements with its PSI partners. The United States has also entered into bilateral ship boarding agreements with Liberia, Panama, the Marshall Islands, Croatia and Belize that set out procedures for obtaining the consent of the flag state on an expedited basis (US State Department Bureau of Nonproliferation, http://www.state.gov/t/np/c10390.htm).

Following the attack on the World Trade Centre in New York on September 11, 2001, the International Maritime Organization (IMO) adopted Assembly Resolution A.944(22) calling for a review of the existing measures and procedures to prevent acts of terrorism which threaten the security of passengers and crews and the safety of ships. The IMO Legal Committee undertook a review of the 1988 Convention for the Suppression of Unlawful Acts Against the Safety of Maritime Navigation (1988 SUA Convention) in order to add new provisions designed to combat the threat of maritime terrorism. After three years of study and deliberation by the Legal Committee, significant revisions were made to the 1988 SUA Convention and its Fixed Platforms Protocol at an International Diplomatic Conference on the Revision of the SUA Treaties organized by the IMO in London from 10-14 October 2005. The revisions were made through the adoption of the 2005 SUA Protocol.

The 2005 SUA Protocol contains two significant changes. First, it broadens the list of offences to include the use of a ship as a weapon and the transport of weapons of mass destruction. Second, it introduces provisions for the boarding of ships in international waters where there are reasonable grounds to suspect that the ship or a person on board the ship is, has been, or is about to be, involved in the commission of an offence under the Convention. The 2005 SUA Protocol contains ship boarding provisions that establish a comprehensive set of procedures and protections designed to facilitate the boarding of a vessel that is suspected of being involved in a SUA offense. The boarding provisions are consistent with existing international law and practice, including the general principle that ships on the high seas are subject to the exclusive jurisdiction of the flag state. Boardings can only be conducted seaward of the outer limits of any state's territorial sea with the express consent of the flag state.

Joint Cooperation Zones could be established in the Indian Ocean Region in areas seaward of the outer limits of the territorial sea by states cooperating in the PSI or by states parties to the 2005 SUA Protocol. In addition, if there are areas in the Indian Ocean where other illegal activities take place which pose a threat to maritime security, interested states could enter into cooperative arrangements to patrol for suspect vessels. Such cooperation could include patrols to prevent attacks on ships exercising passage on international shipping routes in the Indian Ocean. Navies of participating states could also board and search such suspect vessels while they are on the high seas or in an exclusive economic zone, so long as they comply with the principle of flag state consent.

If there are areas of exclusive economic zones where there is a significant problem with illegal activities at sea, such as people smuggling or arms smuggling, interested states could establish a zone of cooperation in the exclusive economic zone, and conduct patrols in the zone to combat these activities. For example, if it were found that weapons were being smuggled from rebels in the Aceh Province of Indonesia to rebel groups in Provinces in southern Thailand, interested states such as Indonesia, Thailand, Malaysia and India could agree to establish a joint cooperation zone to patrol for suspect vessels in the northern approach to the Malacca Strait. Under this arrangement, the partner states could agree in advance to expedited procedures which would enable the naval forces of partner states to obtain expedited consent to board a suspect vessel flying their flag. The arrangement could include provisions on the extradition of persons arrested for engaging in illegal activities in the zone.

Another area in Southeast Asia where a joint cooperation zone might be considered is the Sulu-Sulawesi Sea Area. Not only have smuggling, piracy and other international criminal activities taken place in the Sulu Sea, but the Sulu-Sulawesi Sea area is one of the richest marine areas in the world in terms of biological diversity, and proposals have been made to establish a marine eco-region in the area (see http://www.ssme-wwf.net/index.php). Cooperation among the neighbouring states of Indonesia, Malaysia and the Philippines is essential to combat illegal activities and to protect and preserve the marine environment in this area. A joint cooperation zone would provide a vehicle for the necessary cooperation. There may be similar areas in other parts of the Indian Ocean where joint cooperation zones could be established to enhance maritime security.

Joint Cooperation Zones in Areas Subject to Sovereignty of the Coastal State

Cooperation to establish joint cooperation zones in areas subject to the sovereignty of coastal states is more difficult to achieve. Coastal states have sovereignty in their territorial sea and in their archipelagic waters, as well as in those parts of straits used for international navigation that are within their territorial sea or archipelagic waters. Although the sovereignty of the coastal state in these waters is subject to the regimes of innocent passage, transit passage and archipelagic sea lanes passage, no state may undertake patrols or other security measures in such waters without the consent of the coastal state. It is a fundamental principle of international law that no state can exercise police power in areas under the sovereignty of another state without the consent of the other state.

Many coastal states in Asia (and elsewhere) are very sensitive about their sovereignty, and vigorously oppose any proposals for joint patrols or zones of cooperation that could undermine or infringe their sovereignty. This is to be expected. Even if one starts with the premise that the sovereignty of the coastal state in its territorial sea or archipelagic waters cannot be compromised, it is still possible to establish zones of cooperation to enhance maritime security. If a coastal state is unable to fulfill its obligation to keep its waters safe and secure for foreign ships because of a lack of resources, it should be willing to cooperate with other states who are willing and able to assist it, provided that the arrangement does not infringe its sovereignty. In addition, joint cooperation zones could be a useful tool to enable user states to share more of the burden of ensuring maritime security in sea lanes of communication. However, for such an arrangement to be established, the participating states would have to agree to abide by principles that would ensure that the policing activities of the foreign vessels would not infringe on the sovereignty of the coastal state. The following nine principles could be the starting point for negotiations between littoral states and partner states:

1. The littoral state should formally request assistance from the partner state to assist it in carrying out police powers in waters under its sovereignty under a joint partol arrangment in a zone of cooperation.

2. The partner state should agree that any patrol vessels and crews it provides under this arrangment would be under the general supervision and control of the designated authority of the littoral state.

3. The partner states should agree to coordinate activities of their patrol vessels with the patrol vessels of the littoral state.

4. The partner state should agree that it will confine its patrols to the agreed zones of cooperation.

5. The partner state should agree that its patrol vessels will remain in constant radio contact with the local commander.

6. The partner state should agree that it will allow an officer from the littoral state to be on board its patrol vessel as an observer when its patrol vessel is operating in the waters of the littoral state.

7. The partner state should agree that its patrol vessels will operate under the rules and procedures established by the littoral state for the boarding and search of suspect vessels within its waters.

8. The partner state should agree that it will turn over to the littoral state any persons arrested as a result of the boarding and search of a vessel in their waters, as well as any property seized during the boarding and search.

9. The partner state should agree that its patrol vessels will engage in joint training exercises with the forces of the littoral state.

If these nine principles are followed it should be possible for partner states to assist in securing the sea lanes of communication within waters under territorial sovereignty without infringing on the sovereignty of the littoral states. The exact legal arrangement would have to be carefully negotiated, and may have to be modified in order to comply with the domestic laws of each partner state.

Issues of state responsibility would also have to be addressed. The rule of state responsibility that would apply is Article 6 of the International Law Commission's Draft Articles on Responsibility of States for Internationally Wrongful Acts, 1991, which provides:

The conduct of an organ placed at the disposal of a State by another State shall be considered an act of the former State under international law if the organ is acting in the exercise of elements of the governmental authority of the State at whose disposal it is placed.

The official commentary to the article states that the notion of an organ "placed at the disposal of" the receiving state is a specialised one, implying that the organ is acting with the consent, under the authority and for the purposes of the receiving state.

Zone of Cooperation in the Straits of Malacca and Singapore

The Straits of Malacca and Singapore pose a particularly difficult problem with regard to maritime security because the southern half of the Malacca Strait (from Port Klang south to Singapore) and the entire Singapore Strait are within the territorial sea or archipelagic waters of the littoral states.

Indonesia, Malaysia and Singapore have a long history of cooperation to enhance navigational safety and security in the Straits of Malacca and Singapore. In response to the attacks on ships in the Straits, they have cooperated by the establishment of "coordinated patrols". Coordinated patrols recognise the sovereignty of each state in their territorial sea. Each state is responsible for patrolling within its own territorial sea, but the patrol partners remain in contact with each other and coordinate their activities. The states have not agreed to "joint patrols" in which there may be a single command structure patrolling designated waters of two states with vessels from each state.

In July 2004, Indonesia, Malaysia and Singapore Armed Forces announced a new regional maritime security initiative for the Singapore and Malacca Straits called the Trilateral Coordinated Patrols Malacca Straits, code-named MALSINDO. MALSINDO was announced as a "trilateral coordinated patrol". One major difference was to enhance the number of coordinated patrols. Whereas previously coordinated patrols took place several times a year between the patrol partners on a bilateral basis, under MALSINDO the three states agreed to conduct patrols throughout the year. It has been reported that as part of the operation, each navy is committed to providing between five and seven ships to patrol the Malacca Strait, and that they have established a hotline that will allow them to communicate so as to better coordinate the operation, particularly when a vessel from one of the countries is in pursuit of pirates. In addition, the Jakarta Post reported on 20 July 2004 that under MALSINDO, a warship from one country will also be allowed to enter the waters of another country when chasing a pirate ship, provided that this is communicated first to the host country. If this report is correct, this "right of hot pursuit" may be a major change from previous arrangments for coordinated patrols, and could make a significant difference. Such an arrangement would be a modification of the general rule on hot pursuit under 1982 UNCLOS, which provides that the right of hot pursuit ends as soon as the ship being pursued enters the territorial sea of its own or of a third state (Article 111).

The Malacca Strait is particularly dangerous for international shipping. To date, any offers by third states to assist the littoral states in patrolling the strait have been rejected out of hand because patrols by foreign vessels are viewed by the littoral states as a direct threat to their sovereignty. However, if one of the littoral states were to specifically request a partner state to assist in patrolling its waters in the Malacca Strait in order to prevent illegal activities, including attacks on ships exercising the right of transit passage, such a request would be an exercise of the sovereignty of the littoral state. If arrangements were negotiated based on the nine principles outlined earlier, this would not be an infringement on the sovereignty of the littoral state. Rather, it would be a specific example of "burden sharing" by a user state to cooperate with a littoral state in maintaining the safety and security of maritime navigation.

Authority for such burden-sharing arrangements in straits used for international navigation is set out in Article 43 of 1982 UNCLOS, which reads as follows:

> Article 43. Navigational and safety aids and other improvements and the prevention, reduction and control of pollution User States and States bordering a strait should by agreement cooperate:
>
> (a) in the establishment and maintenance in a strait of necessary navigational and safety aids or other improvements in aid of international navigation; and
> (b) for the prevention, reduction and control of pollution from ships.

Although Article 43 does not specifically mention maritime security, the practice of the International Maritime Organization (IMO) since the attack on the World Trade Centre on September 11, 2001, has been to incorporate security within navigational safety. Therefore, agreements to enhance security of shipping on sea lanes of communication in straits used for international navigation would be agreements within Article 43.

User states benefit more from the Straits of Malacca and Singapore than the littoral states. Nevertheless, the littoral states have the obligation to keep the strait secure and safe for international navigation. Various types of agreements are possible between user states and littoral states under Article 43. An agreement to establish a joint cooperation zone within the Straits to enhance the safety of the sea lanes would be such an agreement. The littoral state would benefit from such an agreement because it would receive assistance in fulfilling its international obligations without any infringement on its sovereignty. The international community would benefit because it would enhance the safety of vessels exercising transit passage through one of the most important routes used for international navigation.

Another option that might be considered is to confine the joint cooperation zone to the traffic separation scheme. The traffic separation scheme extends from Port Klang in the Malacca Strait to the eastern end of the Singapore Strait. The entire traffic separation scheme is within the territorial sea or archipelagic waters of Indonesia, the territorial sea of Malaysia or the territorial sea of Singapore. The international community has the greatest interest in ensuring the safety and security of ships exercising the right of transit passage through the Straits within the traffic separation system. Therefore, it makes sense for the foreign patrol vessels to assist the littoral states in patrolling these waters in order to more equitably share the burden of ensuring safe passage through the Straits. If the joint cooperation zone were established in accordance with the nine principles set out above, the foreign patrol vessels would not pose any threat to the sovereignty of the littoral states. To effectively implement a joint cooperation zone in the Straits, foreign patrol vessels would have to operate in the waters of the littoral states. Therefore, it might be necessary to establish joint patrols in the joint cooperation zone. Such joint patrols should be under the command of an officer of one of the littoral states. The command of joint patrols could rotate between Indonesia and Malaysia in the Malacca Strait, and between Indonesia and Singapore in the Singapore Strait.

Conclusion

It is possible to establish zones of cooperation to enhance the safety of vessels in the sea lanes of communication in a manner that is consistent with the international law of the sea. In particular, it is possible to establish zones of cooperation in the Straits of Malacca and Singapore which would enable foreign patrol vessels to assist littoral states in fulfilling their obligation to keep their waters safe and secure for international shipping. If partner states are willing to attempt to find creative solutions to achieve the common objective of enhancing maritime security in sea lanes of communication, it is possible to do so. Joint cooperation zones are one possible option, and should be considered.

Chapter 17

Multiples Uses, Maritime Law Enforcement and the
Role of Ports and Coast Guards

Sam Bateman

Introduction

Multiple use management is an approach to ocean management that aims to achieve integration among different activities at sea and an acceptable balance of outcomes across the full range of ocean uses. This reflects the principle in the Preamble to the 1982 UN Convention on the Law of the Sea (UNCLOS) that "the problems of ocean space are closely interrelated and need to be considered as a whole" (United Nations, 1983, 1). It has been translated into the well-established concept of integrated oceans management to be adopted both at the national and regional levels. This became an explicit principle, for example, in the Seoul Ocean Declaration adopted for the Pacific Ocean by the Asia Pacific Economic Cooperation (APEC) Ocean-related Ministers in Seoul in 2002.

This chapter argues that this approach applies as much to maritime law enforcement as it does to other areas of oceans management. A multiple use approach to maritime law enforcement comprehends the full range of illegal activities at sea. It uses an integrated approach to maintaining maritime law and order based on agreed international regimes, cooperation, situational awareness, and information sharing. This approach calls for cooperation among all the agencies involved in maritime law enforcement both at the national and regional levels. Of the many institutions involved in maintaining comprehensive maritime security, ports and coast guards have an important role to play, but it is essential that their activities are coordinated in a way that achieves the most efficient outcomes and avoids duplication of effort.

The 1998 Report by the Independent World Commission on the Oceans, The Ocean-Our Future, emphasised the importance of the oceans to the future of the world. It pointed out the oceans are the setting of major problems: territorial disputes that threaten peace and security, global climate change, illegal fishing, habitat destruction, species extinction, pollution, drug smuggling, congested shipping lanes, sub-standard ships, illegal migration, piracy and the disruption of coastal communities are among the problems confronting the international community.

The Ocean-Our Future proposed a comprehensive and integrated approach to addressing these problems of the oceans and providing maritime security. It called for the active involvement of navies in law enforcement in the face of new threats - such as those emanating from illegal trade, clandestine transport of persons, eco-crime, congested shipping lanes, piracy, and terrorism - to make the oceans safer for the global community (Independent World Commission on the Oceans, 1998, pp. 17). However, navies seem reluctant to play this role. They are increasingly preoccupied with their war-fighting role and more traditional forms of security. They are organised and equipped to fight wars and combat military threats, while coast guards are primarily concerned with social, resources and environmental threats to national and global well-being. Because they deal routinely with threats such as pollution and illegal fishing, coast guards are thus proving to be more comfortable than navies with a multiple use approach to maritime law enforcement. In the longer-term, some form of cooperative international "Ocean Guard" might be established to provide maritime law enforcement against multiple use threats.

Good Order at Sea

Good order at sea permits the free flow of seaborne trade and ensures that states can pursue their maritime interests and develop their marine resources in an ecologically sustainable and peaceful manner in accordance with international law. It provides for the preservation and protection of the marine environment, including the conservation of species, and ensures that all states and peoples, including future generations, benefit equitably from the marine environment and the exploitation of its resources.

A breakdown in good order at sea is evident with unlawful pollution of the marine environment, unregulated or illegal fishing (including the use of explosives or chemicals to catch fish), or if other illegal activity occurs at sea. Illegal activity might include piracy; maritime terrorism; maritime theft and fraud; human smuggling; and the shipment of drugs, arms, protected animal and plant species, certain toxic materials and nuclear waste, as well as the dumping of environmentally harmful and hazardous substances banned under international agreements.

Countries may face difficulties in combating illegal activities at sea due to a shortage of trained personnel, the lack of modern equipment, the obsolescence or inadequacy of much national legislation, and the weak maritime law enforcement capability of national agencies. Even developed countries with sophisticated maritime patrol and surveillance capabilities may have difficulty in adequately policing expanded offshore areas.

The escalation and global reach of organised crime has affected all modes of transport, especially maritime transport. Maritime transport constitutes a preferred mode for smuggling illicit goods, such as arms, narcotic drugs, and persons from one country to another, since it may be less detectable than other methods and permits the shipment of large quantities in one consignment. While direct voyages of vessels from countries of origin or supply to receiving countries are primary threats, circuitous routes may be used, involving transshipment at some intermediate port.

The Roles of Maritime Security Forces

The maritime security forces of any country may comprise a navy and a separate para-military force, such as marine police or a coast guard, to undertake constabulary tasks, particularly those of a civil policing nature. This chapter uses the term "coast guard" to refer to these latter forces, although they may have different names in different countries - for example, the Malaysian Maritime Enforcement Agency.

Based on a categorisation originally provided by Ken Booth, maritime security forces are required to fulfill three main functions or roles (Booth, 1977, especially pp. 15-16). These are:

> A military or war-fighting role to defend the state against threats, primarily of a military nature. This role requires capabilities for combat operations either at sea (e.g. surface warfare, anti-air warfare, submarine warfare, maritime strike, mine warfare, protection of shipping or coastal defence), or from the sea (e.g. amphibious operations, naval gunfire support, or land strike using cruise missiles). These may be either sea assertion operations to assert the ability to use the sea for one's own purposes, including for power projection, or sea denial operations to deny the use of the sea to an adversary.

> A constabulary or policing role concerned with sovereignty protection and the enforcement of the full range of national laws at sea. Specific tasks might include maritime surveillance and enforcement, sea patrol, fisheries protection, search and rescue (SAR), combating drug smuggling and piracy, mitigating ship-sourced marine pollution, and controlling illegal immigration. The importance of this role has increased for most countries in recent years due to higher levels of offshore resource exploitation, greater marine environmental awareness and increased illegal activity at sea (United Nations, 2002, pp. 26).

A diplomatic role involving the use of security forces as instruments of foreign policy. This may range from straightforward, rarely controversial activities to support foreign policy objectives, such as civil assistance and regional security cooperation (e.g. port visits, personnel exchanges, passage exercises, etc) through to manipulative/coercive naval presence missions to influence the political calculations of other states in situations short of actual conflict. These latter missions are in line with Sir James Cable's "gunboat diplomacy" (1981), or what Edward Luttwak (1974, especially chapter 1) has called "naval suasion".

In the past, the diplomatic role has been very much the prerogative of navies, while coast guards, to the extent that they existed, were only employed in home waters. This situation has changed. As demonstrated, for example, by recent operations of the Japan Coast Guard (JCG) in Southeast Asian waters to assist the region in its efforts to counter piracy and maritime terrorism (Bradford, 2004), coast guards are being used increasingly as instruments of foreign policy in waters well beyond the limits of national jurisdiction. Some regional countries are also demonstrating a preference for deploying coast guard ships and personnel in sensitive situations at sea rather than naval ships and personnel. Coast Guards are thus emerging as more important national institutions with the potential to make a major contribution to regional oceans governance and security cooperation.

While coastal states have greater rights in their littoral waters under UNCLOS, they also have increased responsibilities. A coastal state has to maintain safety in its waters, protect the marine environment, and generally maintain good order at sea. Surveillance and enforcement operations to protect rights and fulfill obligations at sea require capabilities to monitor activities ("surveillance"); and intercept, board, inspect, and, if necessary, detain or arrest vessels or individuals believed to be acting illegally ("enforcement"). Surveillance may be undertaken by aircraft, satellite or land-based radar, but enforcement invariably requires a surface vessel to intercept, and, if necessary, board, search and arrest a suspect vessel.

The Emergence of Regional Coast Guards

Trends towards the establishment or further development of coast guards are a clear consequence of the recognition by states of the significance of national rights and obligations under the new law of the sea. Table 1 shows how countries in the Indian Ocean Region are approaching the allocation of responsibility for maritime policing or constabulary tasks between navies and separate para-military forces. A preference for a separate service is evident, with 16 of the 23 countries in the survey opting for a para-military service to undertake maritime enforcement although the smaller island countries in particular (Maldives, Mauritius and the Seychelles) only have a coast guard. Some of these new para-military services are very recent and are still in the process of being established (for example, in Indonesia and Malaysia). With the notable exception of France, generally only the smaller countries have not established a separate service. Some countries, like France, have a navy but no coast guard (for example, Iran, Tanzania and Sri Lanka).

Table 1
Maritime Security Forces - Indian Ocean and Littoral States

Country	Navy	Para-Military Service	Name of Separate Para-Military Service
Australia	Yes	Yes	Australian Customs Service
Bangladesh	Yes	Yes	Coast Guard
France	Yes	No	
India	Yes	Yes	Coast Guard
Indonesia	Yes	Yes	Sea Security Coordination Agency
Iran	Yes	No	
Kenya	Yes	Yes	Police Naval Sqn/Customs
Madagsacar	Yes	Yes	Maritime Police
Malaysia	Yes	Yes	Maritime Enforcement Agency
Maldives	No	Yes	Coast Guard
Mauritius	No	Yes	Coast Guard
Mozambique	Yes	No	
Myanmar	Yes	Yes	People's Pearl and Fishery Ministry
Oman	Yes	Yes	Police Coast Guard
Pakistan	Yes	Yes	Coast Guard
Seychelles	No	Yes	Coast Guard
Singapore	Yes	Yes	Police Coost Guard
Somalia	No	No	
South Africa	Yes	Yes	Police Coast Guard
Sri Lanka	Yes	No	
Tanzania	Yes	Yes	
Thailand	Yes	Yes	Marine Police
Yemen	Yes	Yes	Coast Guard

Main Sources: Jane's Fighting Ships and The Military Balance 2003-2004

There are several reasons for establishing a separate coast guard. Legal considerations are important. A coast guard should be a para-military organization. Its officers must have the ability to enforce national maritime laws with wide powers of arrest over both foreigners and national citizens, but, in many countries, there are constitutional and political reasons why military forces should not be involved in policing duties against national citizens (Smith, 1997). In the United States, for example, the military is constrained by the principle of *posse comitatus* with the *Posse Comitatus Act* embodying the traditional American principle of separating civilian and military authority and prohibits the use of the military in civilian law enforcement.

Coast guard units are more suitable than warships for employment in sensitive areas where there are conflicting claims to maritime jurisdiction and/or political tensions between parties. Warships are high profile symbols of sovereignty whose employment in disputed maritime areas may be provocative. In sensitive areas, the arrest of a foreign vessel by a warship may provoke tension, whereas arrest by a coast guard vessel may be accepted as legitimate law enforcement signaling that the arresting party views the incident as relatively minor. A basic clash also exists between the military ethos of applying maximum available force and that of law enforcement, which is more circumspect and usually involves minimum force.

Maritime safety is often a task assumed by coast guards. This task involves issuing certificates for seafarers, ship inspections, vessel traffic management and accident investigations. It does not sit easily with conventional naval activities. Lastly, there is the issue of costs with coast guard vessels and aircraft generally being less expensive than naval units. Furthermore, in developing countries, the civil nature of the coast guard's role may support access to funding from international aid agencies to acquire new vessels. An example of this process is the acquisition by the Philippines Coast Guard of two large (56m length overall) "search and rescue vessels" that are clearly patrol vessels in every respect other than name (Beecham, 2001).

The last ten years or so have seen major developments with the emergence, evolution and employment of coast guards. These developments have been particularly rapid since about 1998. Bangladesh, the Philippines and Vietnam have all established coast guards and both Malaysia and Indonesia are also establishing Coast Guards.

Since 2000, Japan has been actively using the JCG as a "foot in the water" in Southeast Asian waters, ostensibly to combat piracy (Chanda, 2000; Valencia, 2000). The use of the Japanese Maritime Self Defense Force (JMSDF) would be unacceptable for this activity both for Japan constitutionally and to regional countries, with some opposing the JCG presence at least initially. However, Japan has persisted and JCG vessels and aircraft are now regularly visiting Southeast Asia for exercises with their regional counterparts and for anti-piracy patrols. In a further diplomatic initiative, the JCG sponsored the Indian Ocean (Bangladesh, India, Pakistan and Sri Lanka) Maritime Safety Practitioners' Conference in Tokyo in November 2001 (Japan Coast Guard, 2002, pp. 88).

Some National Arrangements and Developments

Australia

Australia's arrangements for the enforcement of national legislation and the protection of national interests at sea are currently quite complex, not least of all because Australia has a federal system of government and the states and territories have jurisdiction over internal waters and the first three nautical miles of territorial sea. There are twelve Commonwealth agencies with an ongoing role in maritime enforcement and compliance, whilst others may request assistance on an ad hoc basis.

The principal civil maritime surveillance and enforcement agency is Coastwatch, a division of the Australian Customs Service (ACS), with a two-star officer from the Royal Australian Navy (RAN) seconded as its Director-General. It manages the aerial surveillance programme, coordinates surface response operations when required by its "client" agencies, and develops intelligence systems for maritime surveillance and enforcement. RAN patrol boats provide much of the surface response, but there is also a National Marine Unit (NMU) of Australian Customs Vessels (ACVs) under the ACS. The NMU has eight sea-going patrol vessels, the Bay class, based mainly in northern Australia and routinely deployed in the constabulary role. As from November 2004, the NMU also included the ice-strengthened, 105-metre vessel, *Ocean Viking*, for patrolling Australia's maritime zones in the Southern Ocean, and maintained under a two-year lease by P&O Maritime Services.

The NMU and Coastwatch are becoming the de *facto* Australian Coast Guard. Seagoing Australian Customs personnel are now armed, well trained and uniformed. They exhibit all of the characteristics of members of a para-military constabulary force. The ACS and/or Coastwatch routinely represent Australia in regional coast guard meetings.

Bangladesh

Bangladesh formed a coast guard in December 1995 under the Department of Home Security to strengthen the state's capacity in the areas of anti-smuggling, anti-piracy and protection of offshore resources. Two ships were initially made available on loan from the Navy but additional vessels have since been acquired, including decommissioned cutters from the US Coast Guard. The roles of the Bangladesh Coast Guard, as approved by the Bangladeshi Parliament (1994), include "participation in rescue and salvage operations in times of natural catastrophes and salvage of vessels, human beings and goods met with an accident". Since its establishment, the Coast Guard has had a major role in dealing with maritime natural disasters (for example, storm surge, floods, cyclones) and ferry accidents to which Bangladesh is prone.

India

The Indian Coast Guard (ICG) is the principal agency for the enforcement of national laws in the maritime zones of India (Chaudhari, 1997). The ICG was established in 1978 as an independent paramilitary service to function under the Ministry of Defence but with its budget met by the Department of Revenue. However, following recommendations from a Group of Ministers (Government of India, 2001, pp. 71), the ICG was brought into the Defence budget in 2001, although the Director-General of the ICG will report to the Defence Secretary rather than to the Chief of Naval Staff. There have been calls that it should now be brought fully under the Navy on the grounds that the "line between war and peacetime has blurred" with increased ship hijacking, gun running and drug smuggling giving "a new dimension to security threats at sea" (Das, 2002).

The functions of the ICG include the safety and protection of offshore installations, search and rescue, marine environmental protection, anti-smuggling operations and enforcing national laws in the maritime zones. Surveillance and patrol of these zones is the primary raison d'étre of the service (Roy-Chaudhury, 2002, pp. 68). The force levels of the ICG have been gradually increased over the years and it currently operates about 55 vessels, including thirteen large offshore patrol vessels, and 20 aircraft. Of these vessels, about 30 are capable of sustained operations offshore while the others are for inshore operations. It would function in support of the Navy during wartime on tasks such as the defence of offshore installations, local naval defence, examination services, control of merchant shipping, maritime surveillance and support in amphibious operations and logistic support (Sen, 1995, pp. 8). Following the success of the Indian Coast Guard in retaking the pirated Japanese vessel *Alondra Rainbow* in November 1999, India has also been advocating joint action on Asian sea piracy.

Indonesia

As one of the major archipelagic states in the world, Indonesia is very much aware of the extent of its maritime interests and of the need to protect its maritime sovereignty and to maintain law and order at sea. However, its efforts in this regard have been thwarted by both the lack of capacity and of coordination between the various government agencies that have responsibility for some aspect of maritime enforcement. At least ten agencies have been identified as being involved in maritime security management with nine authorised to conduct law enforcement operations at sea (Mangindaan, 2004). The situation has been further complicated since the collapse of the Suharto Government by government reforms, including the enactment of autonomy laws that involve devolution of authority to Provincial governments, including possibly some responsibility for law enforcement at sea (Dirhamsyah, 2005, pp. 9). Other reforms include the separation of the Indonesian National Police from the Armed Forces.

The current reform agenda includes the establishment of an Indonesian Coast Guard - *the Kesatan Penjaga Laut dan Pantai*, abbreviated as KPLP (literally the Indonesian Beach and Coast Guard Unit), under the Ministry of Transportation (Mangindaan, 2004). A Sea Security Coordination Agency, referred to as BAKORKAMLA, has also been established to have oversight of sea security and relevant policy. Hitherto the responsibility for law enforcement at sea has largely rested with the Indonesian Navy, which has had the ships and aircraft but not the legal powers to deal with the range of illegal activities at sea (Urquhart, 2003). The Japan Coast Guard (JCG) is playing a key role in the establishment of the KPLP with the stationing of three JCG officials in the Indonesian Ministry of National Development Planning (BAPPENAS) to assist in planning the new service (Urquhart, 2003).

Malaysia

Malaysia provides a good example of the general trend towards navies concentrating on their war-fighting role and separate coast guards being established. Over a decade ago, Mak Joon Num (1993) described how the Royal Malaysian Navy (RMN), like virtually all the navies of maritime Southeast Asia, was caught on the horns of a dilemma in terms of finding a proper balance between the coast guard function and the war-fighting mission. During the years when the USN provided the overall maritime security umbrella for the Asia-Pacific region and the main threats to national security were perceived as land-based, the RMN was able to concentrate its attention on constabulary missions. However, times have changed, and Mak (1993, pp. 123) has identified a need for the RMN to get back to basics and concentrate on its primary war-fighting mission, and, if need be, hand over its lesser patrol vessels to another agency responsible for coast guard roles.

In line with this approach, Malaysia has moved in recent years to establish a separate coast guard. The legislation to establish this service, the Malaysian Maritime Enforcement Agency (MMEA), was ratified by the Malaysian Senate in June 2004 (Taib, 2004). It became operational in March 2005 and has taken over law enforcement in Malaysia's territorial sea and EEZ (Mahadzir, 2005, pp. 26). Its responsibilities also include search and rescue, pollution control and counter-piracy and drug trafficking on the high seas (Taib, 2004). The impetus to establish the new agency largely came from the RMN whose officers "thought it was a waste of resources to use large, sophisticated warships for maritime enforcement given the wear and tear on extended operations, as well as the time taken up in court appearances by officers aboard ships that made arrests" (Mahadzir, 2005, pp. 26). The Agency will operate at least six helicopters and fixed-wing aircraft and about 80 small and medium-sized vessels. These assets are drawn from diverse maritime agencies, such as the Marine Police, Navy, and the Fisheries and Customs Departments.

Maldives

The Republic of the Maldives provides a good example of the arrangements for maritime law enforcement that are typical of a small island state. Maritime surveillance and enforcement are responsibilities of the Coast Guard, which is part of the National Security Service (NSS). In addition to the Coast Guard, the NSS comprises the Police Force, a Quick Reaction Force (QRF) of three platoons, and Prisons and various Service Units (Engineering, Catering, Fire Services, Transport, Communications, Training, and so on). All members of the NSS undertake common basic training. The NSS Coast Guard either enforces or assists in the enforcement of all national laws applicable to the maritime jurisdiction of the Maldives. The present fleet of the Coast Guard comprises about ten patrol craft up to about 25 metres in length.

Singapore

Formerly known as the Marine Police, the Singapore Police Coast Guard operates a large fleet of more than 80 port security and fast interceptor craft for combating piracy, drug smuggling and illegal entry (Saunders, 2001, pp. 622; Singapore Police Coast Guard, 2005). The Coast Guard works closely with the Republic of Singapore Navy (RSN) and the Indonesian Navy (TNI-AL) to provide coordinated patrols to counter acts of piracy and armed robbery against ships in the Singapore Straits.

Thailand

The Royal Thai Marine Police is part of the Thai Police Bureau. The service acts as a coast guard in inshore waters with about 65 armed patrol craft (the largest of which is about 600 tonnes) and over 65 smaller vessels equipped only with small arms (Saunders, 2001, pp. 694). It has increased commitments to fight terrorism and counter piracy, particularly in the northern part of the Malacca Strait and has been seeking additional funds to upgrade its fleet (Sookpradist, 2003).

Coast Guard Cooperation

Cooperation between coast guards may offer benefits not available with naval cooperation. Warships from major maritime powers may overwhelm vessels from small navies by their sheer size, technology and firepower. On the other hand, coast guard vessels may appear less intimidating and in periods of tension they may be less provocative (Stubbs, 1994, pp. 518). They are "less threatening than larger, more heavily armed haze-gray warships", and are able to carry out exercises and training with other states that might not be possible on a navy-to-navy basis (Truver, 1998, pp. 45). It has been observed, for example, that the Chinese have been "wary" of working closely with the US military for fear that this might reveal their weaknesses, and that while planned USN-Mexican Navy exercises in 1996 met with outrage and political controversy in Mexico, cooperation between the USCG and the Mexican Navy was able to proceed routinely and quietly (Truver, 1998, pp. 45-46).

Cooperation between regional coast guards is expanding rapidly. Under the leadership mainly of the JCG, the Heads of Asian Coast Guards meetings now have some significant achievements. The Asia Maritime Security Initiative 2004 (AMARSECTIVE 2004) was agreed at the meeting in Tokyo in June 2004, and the more recent Regional Cooperation Agreement on Combating Piracy and Armed Robbery against Ships in Asia (ReCAAP) was agreed in November 2004. All ASEAN states, Japan, China, Korea, India, Bangladesh and Sri Lanka are working under ReCAAP to set up an information network and a cooperation regime to prevent piracy and armed robbery against ships in the regional waters. ReCAAP is a very significant achievement for the region that provides the basis for regional cooperation to counter piracy and armed robbery against ships. It provides for the establishment of an Information Sharing Centre (ISC) to be located in Singapore.

The Role Of Ports

Ports have a major role to play in maritime law enforcement, particularly the suppression of piracy and armed attacks against ships. While piratical attacks are committed against ships, offshore installations and their crews, the causal factors and most effective solutions will be found onshore. Patrols at sea by maritime or marine police forces, possibly with the support or direct involvement of regional navies, may be effective in deterring illegal activity at sea but the reality is that very few criminals are actually caught at sea. The most effective solutions for coastal and port states lie in traditional policing methods onshore, including investigation of possible links with organised crime, and in measures to improve the security of ports and harbours.

Port security is an important consideration but the security of a ship is quite different from the security of a port. The former is largely the responsibility of the shipmaster and his crew, except when the ship is in port, and certain other considerations apply, while the latter is entirely the responsibility of the port or coastal State. Security in wharf areas, ports, harbours and anchorages requires active patrolling both ashore and afloat, and the institution of physical measures (that is, effective fencing and access controls, including personal identity documentation) to prevent unauthorised access.

There are two main ways in which ports contribute to maritime security. The first is by conformity with the requirements of the International Ship and Port Facility Security (ISPS) Code developed by the International Maritime Organization (IMO). Following the 9/11 attacks, the IMO has given high priority to the review of existing international legal and technical measures to prevent and suppress terrorist attacks against ships and improve security aboard and ashore. The aim is to reduce the risk to passengers, crews and port personnel both onboard ships and in port areas and to vessels and their cargoes. The ISPS Code, which entered into force on 1 July 2004, includes requirements for ships and port facilities such as the provision of security plans and officers, onboard equipment and arrangements for monitoring and controlling access. The second is by participating in networks of information exchange, particularly those that provide information on the movements of ships and a historical record of incidents of illegal activity. Comprehensive knowledge of what is happening at sea is an essential element of maritime security and ports have an important role in providing that information.

Maritime Domain Awareness

With the War on Terrorism and the priority now accorded by many countries to Homeland Security, a new expression has entered our maritime strategic lexicon. This is *maritime domain awareness* or sometimes, *situational awareness*. Basically this means knowing what is going on in the maritime approaches to the country. What shipping is in the area? What is it doing? Where is it going? What is the cargo? What other maritime activity is out there? It is an integrated approach to maritime security that ties in the threats of maritime terrorism, illegal immigration, drug smuggling, illegal fishing and marine pollution. It suggests the fundamental importance of having good information on which to base risk assessments and establish priorities for maritime security.

As a means of countering the risk of the maritime transportation system being used for terrorist purposes, countries have been introducing systems to provide advanced information on ships about to enter their waters, the crews of these ships and their cargoes. In March 2005, for example, Australia introduced the Maritime Information System, which covers up to 1,000 nautical miles from Australia's coastline. On coming within this distance, vessels proposing to enter Australian ports are required to provide comprehensive information, such as ship identity, crew, cargo, location, course, speed and intended port of arrival. Within Australia's exclusive economic zone (EEZ), the aim is to identify all vessels other than day recreational craft. In a demonstration of an integrated approach to maritime security, the Joint Offshore Protection Command manages the system as a joint command of the Australian Defence Force and the Australian Customs Service.

In another example of the application of maritime domain awareness, the United States proposed the Regional Maritime Security Initiative (RMSI) in May 2004 to improve international cooperation, specifically in the Malacca Straits, against the transnational threats of terrorism, piracy and trafficking. Major elements of RMSI include increased situational awareness, information sharing, a decision-making architecture and interagency cooperation.

The implementation of maritime domain awareness requires the following:

- Comprehensive and specific knowledge of the marine environment, including, for example, the movement of shipping to provide a basis for threat assessments at a regional and national level;

- Less specific knowledge such as extensive knowledge of geography, weather, shipping routes, fishing areas, and so on; and

- An ability to collect, synthesise and analyse all source intelligence, make vulnerability assessments and provide a single, integrated picture of relevant information within the area of interest.

Tracking Ships

In an ideal security conscious world, ships would move around the world like civil aircraft, being passed from one national system of traffic control to another. With initiatives promoted by the United States and now under consideration by the IMO for the Long Range Identification and Tracking (LRIT) of vessels, a system may eventually emerge for commercial ships above a certain size and making use of automatic identification system (AIS) data. However, many other vessels using the world's oceans will remain outside its scope. This inability to monitor the movement of fishing vessels, as well as cruising yachts and other private vessels, will remain a major gap in international arrangements for maritime security.

Even with current LRIT plans there are still unresolved issues. It is by no means certain, for example, that a coastal state has a right to identify and track ships exercising the freedom of navigation either through its exclusive economic zone (EEZ) or on the high seas, and not intending to proceed to a port or an anchorage located within the territory of that coastal state (Hesse and Charalambous, 2004, p. 138). It is also not clear how an LRIT system will deal with the principle of sovereign immunity that applies to warships and government vessels on non-commercial service. Presumably, these vessels will not have to comply with the system. As well as tracking at sea, an effective international system should also include standardised reporting of shipping arrivals and departures but this might arouse both security and commercial sensitivity concerns similar to those now being raised about the free availability of AIS data.

Implementing an Integrated Approach

Implementing an integrated approach to maritime law enforcement at the national level requires processes both for policy development and operational implementation. Policy development requires public sector departments and agencies that have clear divisions of responsibility without duplication of effort. On the operational side, maritime security forces and law enforcement agencies (including the agency responsible for marine search and rescue) also need to avoid overlapping responsibilities and duplication of effort. National arrangements are required for the collection, analysis and dissemination of information to provide situational awareness, and for the determination of risk assessments related to maritime security threats.

National (and State or Provincial) Maritime Security Committees or Maritime Security Task Forces are required to brings together policy and operational agencies and intelligence services. These should be supported by information centres (for example, national focal points, regional coordinating centres, maritime rescue coordination centres, an information sharing centre) to provide the necessary situational awareness of national maritime zones and their maritime approaches.

Maritime security forces should have a cross-sectoral role. It makes no sense to have ships and aircraft of different agencies patrolling in the one area but for different purposes. Coast guards rather than navies tend to able to readily assume an integrated role in maritime law enforcement covering the full range of potential illegal activities at sea and with powers over both national citizens and foreigners. Appropriate national legislation is required that allows the coast guard to take enforcement action without having to refer the circumstances back to another agency.

At a regional level, there is a premium on cooperation and coordination. The geographical situation of the United States, for example, makes it largely independent in building maritime domain awareness. It is able to collect information on its own maritime environment with its own resources and with only a small amount of assistance from its immediate neighbours, Canada and Mexico. The situation is quite different for other countries that are surrounded by neighbours. For these countries, any attempt at gaining awareness of their maritime environment has to be a cooperative endeavour. This is a consequence of both geography and limited national capabilities.

Heads of Asian Coast Guard Agencies Meetings have an important role in building cooperative relations among agencies and establishing arrangements for sharing information on maritime security. The Asian Coast Guard Agencies are the authorities responsible for conducting law enforcement activities of anti-piracy and armed robbery against ships and other unlawful acts at sea, including maritime terrorism, when within their charter and/or providing assistance to persons and/or ships in distress at sea as a result of such attacks. The ReCAAP ISC in Singapore will be an important facility for the collection and analysis of data on piracy and armed robbery against ships.

Looking to the Future

The *Ocean-Our Future Report* believed that navies must assist in meeting the challenge of international oceans management. It noted that: "The role of navies and, where appropriate, other maritime security forces, (should) be reoriented, in conformity with present international law, to enable them to enforce legislation concerning non-military threats that affect security in the oceans, including their ecological aspects" (Independent World Commission on the Oceans, 1998, pp. 17). However, for the reasons discussed in this chapter, it is now unlikely that navies will accept this role and it is more likely to be assumed by coast guards.

There is more and more distance opening up between warships optimised for war-fighting and coast guard type vessels designed for maritime policing. As Colin Gray (2001, pp. 106) has suggested, navies and coast guards are "driven by the beats of different drummers". This is increasingly so with navies being consumed by high technology weapon systems and the concepts of network-centric warfare (NCW) and the revolution in military affairs (RMA). Navies are attracted to larger vessels that can carry more weapons and sensors and are less vulnerable. Even smaller navies such as those of Singapore, Malaysia and Brunei are building larger vessels. Maritime strategists opine that, due primarily to the benefits of networking, "big is beautiful" and smaller numbers of larger vessels have advantages over larger numbers of smaller vessels (Friedman, 2001, pp. 242). However, numbers of hulls remain important for maritime policing tasks.

The type of expertise that navies have is becoming increasingly different to that possessed by coast guards, which, in turn, have expertise that is not the same as naval expertise. Coast guard personnel have to be "lifesavers, guardians and warriors" (Stubbs, 2002). Greater use of the sea, increased illegal activity at sea and concern for the marine environment have increased the number of international regimes that are applicable and have made the business of maritime policing more complex.

Part of this conundrum is associated with changing concepts of security. Navies and warships are designed to fight wars and combat military threats while coast guards and coast guard patrol vessels are primarily concerned with social, resources and environmental threats to national well-being and a comprehensive view of security. Some authors talk about "threats without enemies", and the scope for an "Oceanguard" to protect the oceans and their living resources from environmental stress and the loss of biodiversity (Prins, 1993 and 2000). While the constabulary or policing role at sea has been seen in the past as a national one, it is possible that in the future it will involve international policing on the high seas beyond the limits of national jurisdiction. As Gwyn Prins (2000, pp. 411) has described it, an "Oceanguard would be a high seas equivalent of a coast guard, performing similar missions but in defense of the global commons and in the enforcement of international, not specific, national laws in those areas stated to be, and accepted as being, free of the exercise of sovereign rights".

As well as a global Oceanguard under the auspices of the UN or an UN agency, regional arrangements might be possible. An ASEAN Coast Guard has been suggested (Dela Cruz, 1994), as well as well as a Pacific Islands Ocean Guard that would bring together the scarce resources of the Pacific island countries (PICs) into a cooperative maritime surveillance and enforcement regime (Bateman, 2004). A basic framework of legal regimes and cooperation already exists in the PICs, primarily through the Pacific Island Forum, that might be used as a basis for a regional Ocean Guard. The Ocean Guard would police the region not only for fisheries protection but also for economic, environmental protection, humanitarian and constabulary purposes. In the longer term, we may be thinking of an Oceanguard for the Indian Ocean.

Conclusion

Current trends in globalisation are supporting the importance of sea power in general, and at least in the Indo-Pacific region, both navies and coast guards are growing. However, most navies in the region are showing some reluctance at having greater involvement in maritime law enforcement, and coast guards are being expanded to assist in managing regional oceans and seas and countering increased illegal activity at sea. This seems to be a basic response to national needs and a sense of responsibility with regard to fulfilling a shared responsibility and interest in good order at sea. While much naval planning and thinking is a reaction to the past, the development of coast guards is in many ways a response to future needs and concerns over the problems of the ocean environment.

Implementing a multiple use approach to maritime law enforcement that comprehends the full range of illegal activities at sea and provides security against maritime threats requires both cooperation and good information. Ports and coast guards are important elements in the process of integration, along with all the other institutions and agencies involved in providing comprehensive maritime security.

References

Bangladesh Parliament (1994), *The Coast Guard Act, 1994*, Act No.26 of 1994, Article 7 - Functions of the Armed Force.

Bateman, Sam (2004), 'Developing a Pacific Island Ocean Guard: The Need, The Possibility and The Concept', in *The Eye of the Cyclone: Issues in Pacific Security*, ed., Ivan Molloy, Sippy Downs: Pacific Islands Political Science Association & University of the Sunshine Coast Press, pp. 208-224.

Beecham, Bill (2001), "San Juan' and 'Don Emilio", *Asia Pacific Shipping*, Vol. 1 (4), January, p. 18

Booth, Ken (1977), *Navies and Foreign Policy*, London: Croom Helm.

Bradford, John F. (2004), "Japanese Anti-Piracy Initiatives in Southeast Asia: Policy Formulation and the Coastal State Responses", *Contemporary Southeast Asia*, Institute of Southeast Asian Studies, Singapore. Vol. 26 (3), pp. 480-505.

Cable, Sir James (1981), *Gunboat Diplomacy 1919-1979*, London: Macmillan.

Chanda, Nayan (2000), 'Foot in the Water', *Far Eastern Economic Review*, 9 March, pp. 28-29.

Chaudhari, Captain Vijal IN (1997), 'Management of the Maritime Zones of India', in *Policing Australia's Offshore Zones - Problems and Prospects*, eds., Doug MacKinnon and Dick Sherwood, *Wollongong Papers on Maritime Policy*, No. 9, Centre for Maritime Policy, University of Wollongong, pp. 187-192.

Das, Premvir (2002), 'Coast Guard needs sea change', *The Indian Express* online (11 July) <http://www.indian-express.com/print.php?content_id=5760> (date accessed: 16 July 2002).

Dela Cruz, Captain Gualterio I., PN (1994), 'Time for a NEW Coast Guard', *USN Institute Proceedings*, March, pp. 58-60.

Dirhamsyah, Dirham (2005), 'Maritime Law Enforcement and Compliance in Indonesia: Problems and Recommendations', *Maritime Studies*, 144, September/October, pp. 1-16.

Friedman, Norman (2001), *Seapower as Strategy* - Navies and National Interests,Annapolis: Naval Institute Press.

Government of India, 2001, *Reforming the National Security System,* Recommendations of the Group of Ministers, February.

Gray, Colin S. (2001), 'The Coast Guard and Navy - It's time for a "National Fleet"', *Naval War College Review*, Vol. LIV (3), Summer, pp. 112-138.

Hesse, Hartmut and Nicolaos L. Charalambous (2004), 'New Security Measures for the International Shipping Community', *World Maritime University Journal of Maritime Affairs*, Vol. 3 (2), pp. 123-138.

Independent World Commission on the Oceans (1998), *The Ocean - Our Future*, Cambridge University Press: Cambridge.

Japan Coast Guard (2002), *Annual Report 2002*, Tokyo: Japan Coast Guard,

Luttwak, Edward N. (1974), *The Political Uses of Sea Power*, Baltimore: John Hopkins Press.

Mahadzir, Dzirhan (2005), 'New maritime agency steps up', *Asia-Pacific Defence Reporter*, February.

Mak, Joon Num (1993), 'Malaysia's Naval and Strategic Priorities: Charting a New Course', in *Maritime Change - Issues for Asia*, eds., Ross Babbage and Sam Bateman, North Sydney: Allen & Unwin, pp.117-125.

Mangindaan, RADM (ret) Robert (2004), 'The Indonesian Coast Guard: Guarding Coast and Territorial Waters', Paper presented at the First Meeting of the CSCAP Study Group on Capacity Building for Maritime Security Cooperation, Kunming, China, 6-8 December.

Prins, Gwyn, ed., (1993), *Threats without Enemies: Facing Environmental Insecurity*, London: Earthscan.

Prins, Gwyn, (2000), 'Oceanguard: The Need, the Possibility, and the Concept', *Ocean Yearbook 14*, Chicago: Chicago University Press, pp. 398-419.

Roy-Chaudhury, Rahul (2000), *India's Maritime Security*, Delhi: Knowledge World.

Saunders, Commodore Stephen (ed), 2001, *Jane's Fighting Ships 2001-2002*, Coulsdon: Jane's Information Group.

Sen, Pulak (1995), 'India's Coast Guard comes of age', *Maritime International*, Vol.1 (2), February.

Singapore Police Coast Guard (2005), 'Background', Police Coast Guard homepage available at: <http://www.spf.gov.sg/about spf/pcg2/pcgback.htm> (date accessed: 30 March 2005).

Smith, Hugh (1997), 'The Use of Armed Forces in Law Enforcement', in MacKinnon and Sherwood, *Policing Australia's Offshore Zones*, pp. 74-97.

Sookpradist, Kitisom (2003), 'Thai Police Plan to Spend More in FY2004', *Bangkok Post online*, 25 April. As reported on website at: <http://strategis.ic.ca/epic/internet/inimr-ri.nsf/en/gr115943e.html> (date accessed: 31 March 2005).

Stubbs, Captain Bruce B. USCG (1994), 'The US Coast Guard - A unique instrument of US national security", *Marine Policy*, Vol. 18 (6), pp. 506-520.

Stubbs, Captain Bruce B. (2002), 'We are lifesavers, guardians, and warriors', *USN Institute Proceedings*, April, pp. 50-53.

Mat Taib Yasin, (2004), 'The Malaysian Maritime Enforcement Agency (MMEA)', Paper presented at *First Meeting of the CSCAP Study Group on Capacity Building for Maritime Security Cooperation in the Asia-Pacific*, Kunming, China, 6-8 December.

Truver, Scott C. (1998), 'The World is Our Coastline', *USN Institute Proceedings*, June, pp. 45-47

United Nations, (1983), *The Law of the Sea. Official Text of the United Nations Convention on the Law of the Sea with Annexes and Index*, United Nations: New York.

United Nations, (2002), The 2002 Annual Report of the UN Secretary-General on Oceans and the law of the sea noted that in the twenty years since the adoption of UNCLOS, crimes at sea have become more prevalent and are increasing. UN General Assembly, *Oceans and the law of the sea - Report of the Secretary General* dated 7 March 2002 (UN document A/57/57), United Nations: New York.

Urquhart, D. (2003), "Japan helping Indonesia to set up coast guard", *The Business Times online* (25 March) <http://business times.asial.com.sg/sub/shippingtimes/story/ 0,4574,76347,00.html?> (date accessed: 28 March 2003).

Valencia, Mark (2000), 'Joining Up With Japan to Patrol Asian Waters', *International Herald Tribune*, 28 April.

Chapter 18

Securing Sea Lanes of Communication in the Indian Ocean: The Way Ahead

Sanjay Chaturvedi Mat Taib Yasin and Dennis Rumley

The sea is presented as a space "outside" society, a friction-free surface increasingly made obsolete (or at least suitable for being ignored) thanks to air travel and seamless container transport. The sea is presented as an abstract point on a grid, to be developed. The sea is presented as a repository of fragile, global nature, to be stewarded. *Each of these images is accurate, but each is partial.* Not only do these images tell partial stories; they obscure the material reality experienced by those who derive their living from the sea. *These images obscure contemporary seafarers, dockworkers, artisanal fishers, and others who may be "managed" out of existence by the regulatory strategies with which each image is aligned.* For the sea remains - as it has been since the advent of the modern era - a space constructed amidst competing interests and priorities, and it will continue to be transformed amidst social change. (emphasis given)

Philip E. Steinberg (2001: 207)

In conventional wisdom, a sea lane is composed of a neat combination of points and lines at sea which connect ports all around the globe. Especially in view of remarkable advances in maritime transportation technologies, the sea lanes of communication are now better approached and analyzed as a 'maritime highway' embedded in globalizing geopolitical economy. This organically interconnected maritime highway is served and sustained by a feeder services network that links regional ports, land and air transport systems with a central focus on hub ports (the biggest hub ports in the Eurasian Maritime World include Hong Kong, Singapore, Shanghai, ShenZhen, Pusan and Kaohsiung) handling giant containers and mutually connected by a hub and spokes network with a central focus on each hub port. As pointed out by Kazumine Akimoto (2006a), "the term "Consolidated Ocean Web of Communication (COWOC) would perhaps be more appropriate to describe this new world". Moreover, since ancient times, the ocean flowing along the southern and eastern borders of the Eurasian continent, which connects the Indian Ocean, Southeast Asian Seas and East Pacific Ocean (Eurasian Blue Belt) has functioned as the "Silk Road of the Sea" and remains the main oceanic artery supporting the global economy (Ibid.).

The proverbial billion-dollar question, being touched upon, in one form or the other by various contributions to this volume is not whether 'securing' the sea lanes of communications in the Indian Ocean (or for that matter in other parts of the globe) is a critical issue shared by all concerned (it certainly is and will remain so) but how can this 'common' concern be translated into viable practices based on a dialogic consensus among various stakeholders despite diverse images that they might entertain about ocean-space. Different entry points into the subject notwithstanding, one of the key points underlined by various authors in this collection appears to be that the imperative of comprehensive security of the sea lanes can not be divorced from the way(s) in which the Indian Ocean-space has been (and continues to be) socially constructed by various stakeholders. The manner in which diverse geographies of the ocean-space are being written and upheld in the first place also carries enormous implications for how the challenge of securing the sea lanes is being taken up by policy-makers.

In our introductory chapter to this book, while listing out various reasons for the lack of an 'Indian Ocean Regional Maritime Security Regime' we had chosen to highlight, among other important factors, the prevailing inability on the part of all states concerned to view problems holistically, along with a lack of individual and collective will. This could well be due to the fact that, as Philip Steinberg has argued (2001: 207), image-formation plays a crucial role in constituting the framework through which we encounter and interpret "reality". Moreover, behind each image formation there is always a process of production; a social construction. Since uses, regulations, and representations impact and reproduce each other to form a particular social construction of ocean-space, a critical analysis of the processes behind the images of socially constructed Indian-Ocean space(s) demands and deserves utmost attention. As Steinberg (Ibid.) puts it so succinctly,

> Analysis of the processes behind the images of socially constructed spaces is crucial, in part because every image suggests a social policy. If the ocean is a friction-free transport surface outside society, then it should be managed with as few rules as possible to allow for maximum utilization. If the ocean is a set of developed places, then these places should be captured by sovereign states so that political authorities can institute regulations that will facilitate development through the placement of spatially fixed investments. And if the ocean is a resource-space too valuable to be squandered away by competing states and their enterprises, then it should be stewarded (whether by individual states, the community of states, or civil society) so that all nations (or individuals) may be given the opportunity to benefit from its riches.

Government of India, 2001, *Reforming the National Security System*, Recommendations of the Group of Ministers, February.

Gray, Colin S. (2001), 'The Coast Guard and Navy - It's time for a "National Fleet"', *Naval War College Review*, Vol. LIV (3), Summer, pp. 112-138.

Hesse, Hartmut and Nicolaos L. Charalambous (2004), 'New Security Measures for the International Shipping Community', *World Maritime University Journal of Maritime Affairs*, Vol. 3 (2), pp. 123-138.

Independent World Commission on the Oceans (1998), *The Ocean - Our Future*, Cambridge University Press: Cambridge.

Japan Coast Guard (2002), *Annual Report 2002*, Tokyo: Japan Coast Guard.

Luttwak, Edward N. (1974), *The Political Uses of Sea Power*, Baltimore: John Hopkins Press.

Mahadzir, Dzirhan (2005), 'New maritime agency steps up', *Asia-Pacific Defence Reporter*, February.

Mak, Joon Num (1993), 'Malaysia's Naval and Strategic Priorities: Charting a New Course', in *Maritime Change - Issues for Asia*, eds., Ross Babbage and Sam Bateman, North Sydney: Allen & Unwin, pp.117-125.

Mangindaan, RADM (ret) Robert (2004), 'The Indonesian Coast Guard: Guarding Coast and Territorial Waters', Paper presented at the First Meeting of the CSCAP Study Group on Capacity Building for Maritime Security Cooperation, Kunming, China, 6-8 December.

Prins, Gwyn, ed., (1993), *Threats without Enemies: Facing Environmental Insecurity*, London: Earthscan.

Prins, Gwyn, (2000), 'Oceanguard: The Need, the Possibility, and the Concept', *Ocean Yearbook 14*, Chicago: Chicago University Press, pp. 398-419.

Roy-Chaudhury, Rahul (2000), *India's Maritime Security*, Delhi: Knowledge World.

Saunders, Commodore Stephen (ed), 2001, *Jane's Fighting Ships 2001-2002*, Coulsdon: Jane's Information Group.

the organization of political or economic power within them. In other words, "while the sea was present in Indian Ocean social life, it was constructed as a special space of trade, external to society and social processes" R. J. Barendse (2002:45). Moreover, since the ocean was conceived and constructed as "different" from social space, "few if any attempts were made to use ocean-space for projecting power" (Ibid.). The ocean was perceived as a distance and not as a territory. For most of the mercantilist era, the world ocean-space was perceived by and large not as a space to be possessed but as a space in which state-actors exerted power and pursued a degree of control. In short, Steinberg's argument is that throughout this period the territory of ocean-space *per se* elicited little interest.

While no attempts were made by the Indian Ocean littoral states to use ocean space for projecting power (unlike the present times) one could identify a distinctive Indian Ocean maritime world, which, having evolved over millennia through maritime-based commercial and cultural linkages, was remarkably self-contained until the eighteenth century. As Kenneth McPherson (2004: 264) puts it, "maritime highways were an extension of land-based and riverine highways, with each linked to the other by myriad ports and market-places where cargoes of goods and ideas mingled. Long-distance trade, by land and sea, was a cornerstone of international relations in the pre-modern period".

Writing on the 17th century 'Arabian Seas', R. J. Barendse (2002) draws our attention to how the Western Indian Ocean (described by Barendse as the historic core area of the Indian Ocean) of those days could be perceived "not so much as a closely interlocking or unipolar system, like the modern world economy, but as a much looser series of networks; more like a social system that had boundaries, structures, member groups, rules of legitimation and coherence with ties over a wide area. But the commercial nodes of this system penetrated the rural economy rather marginally. What the system also lacked was a central location where prices throughout the system were fixed, (like the City of London or like the Wall Street these days) and a single hegemonic centre (similar to Britain in the 19th century and the United States in much of the twentieth). Yet,

> Commercial ties ranged far and wide - far wider, in fact, than is commonly recognized in most studies of Asian trade, linking areas seemingly as remote from the world market as the interior of Rajasthan or Malawi to a flourishing trade centered on the Arabian Seas and extending beyond it. For the Arabian seas were linked by far-flung trading networks that extended into Europe, Central Asia, East Asia, Southeast Asia, Central Africa and the Americas (Ibid.: 3-4).

To return to Steinberg's interesting thesis, once the transition to industrial capitalism was under way in the mid-18th century, a different spatial logic was introduced. Both classical and neo-classical economic theories idealized the ocean as the anti-thesis of land-space. While propounding the thesis that all nations would eventually attain a maximum level of propinquity and prosperity, provided they were to trade freely with each other, these doctrines idealized the ocean-space as an empty transportation surface (or as an empty void outside society), beyond the realm and reach of society and social relations. But the fact of the matter was that the 'capitalist-era ocean' was indeed a space intensely used (and hence shaped) by society. Hence, the corresponding growing perception, bordering urgency, that some regulation was needed. Whereas, at the same time, the projection of power (especially sea power) in the deep sea was perceived as legitimate only when deployed towards the objective of eliminating obstacles to free navigation.

The point we wish to flag here is that it is often a particular constellation of political-economy, marked and sustained by a dominant spatiality, that gives rise to a social construction of ocean-space, which in turn demands a particular kind of legal 'regulation' and/or 'freedom'. In other words, the interplay between mobility and territoriality is not as simple and straightforward as is often assumed. It is no surprise, therefore, that during the industrial-capitalist era the policy-makers sought a regulatory mechanism for ocean-space in general, which would compliment the 'great void' representation and avoid at the same time a situation where states might start undermining social construction of the ocean as a friction-free transportation surface with its primary resource, namely 'connection'. While advancing towards this objective, policy-makers during this era gradually turned away from the dominant mercantilist period The Freitian model of stewardship (wherein individual states claimed a degree of territorial authority over specific ocean routes) and chose instead a more Grotian model, wherein the sea's resources were stewarded by the global community as a whole. It is important to note that these Grotian instruments generally were implemented in a rather "weak" form, due to the apprehension that stewardship by any entity (even the world community as a whole) might result in certain claims of territoriality that could seriously undermine the great void ideal. Likewise, powers of stewardship often were given to non-state entities rather than states, under the assumption that the former would be less likely than the latter to turn rights of stewardship into assertions of territorial enclosure (Steinberg 1999: 125).

We might also note in passing that as the industrial era progressed and ocean-space increasingly supplied fish and mineral resources, what ensued in the first half of the 17th century (with Britain and Netherlands joining Spain and Portugal as global ocean powers) was the debate between the English jurist John Selden and the Dutch jurist Hugo Grotius. Whereas the former argued in favour of states having the right to enclose and claim discrete areas of ocean space, the latter took the stand that the ocean must remain open to all (Smith and Roach 1997: 283). Skipping the finer points of this interesting debate, we might note in brief that what was rather innovative about Grotius' position was his contention that the holder of stewardship be not one state (as it was in Roman times when Rome claimed rights in the Mediterranean under the claim to imperium but not under the claim to dominium) or individual states in their respective spheres of influence (as it was with Spain's and Portugal's claims to exclusive rights under the 1494 Treaty of Tordesillas system) but rather the 'community' of states (Borgese 1998: 119). In the doctrine of the 'freedom of seas', as propounded by Grotius, one could find an attempt at "unbundling" of territoriality (Ibid.).

Grotius did leave some room in his proposed scheme of things for the individual states to police discrete areas of the sea but only as stewards. By this he implied that these states would remain under an obligation to facilitate the 'basic human right of navigation' and fishing. In contrast to Grotius, Selden envisaged a system whereby coastal states could exercise effective territorial possession in a specific ocean-space. But similar to Grotius, Selden maintained that state governance in this ocean-space was not to jeopardize the 'natural' right to navigation. It is worth noting that the ocean-governance system that prevailed through much of the modern era combined elements of the Roman model of stewardship with that proposed by Grotius and Selden (Steinberg 1999). Moreover,

> Although Britain was the overwhelming sea power for much of the eighteenth and nineteenth centuries, it never sought to claim the world ocean as part of imperial territory. Britain's empire differed from that of Rome in that Britain - hegemonic as it may have been - existed within a political system of formally equivalent sovereigns. Assertions of power that gave even the appearance of claiming the bulk of the sea as British territory likely would have led to counterclaims by other states, which would have destroyed the "freedom of the seas" that was both a goal and a basis of British hegemony. Thus Britain largely chose to operate within a Groatian construct in which the ocean was stewarded by the community of states of which Britain was the dominant constituent member. For instance, in its attempt to ban the slave trade, abolish piracy, regulate maritime transport, and protect undersea cables, the community of states, under British leadership, took pains to avoid strategies that implied system of territorial divisions or individual state control that could interfere with the construction of the ocean as a friction-free space of connection.

By the mid-19th century this Grotian governance system was joined by a Seldian one in which narrow coastal strips were permitted to be incorporated within territory of individual states, but with the proviso that this possession of ocean space be limited by respect for the natural right of all individuals to navigate across the entirety of the ocean's surface (Ibid.: 261). (emphasis added)

The modern history of the law of the sea in general, and the 1982 United Nations Convention on the Law of the Sea (UNCLOS III) in particular, are revealing examples of how various competing modern era social constructions of ocean-space have been sought, negotiated and finally arrived at through a delicately balanced 'compromise package' of legal formulations. UNCLOS III establishes a comprehensive regime for the law of the sea by incorporating the interests of both developed and developing maritime states. One of the sharpest disagreements between the developing coastal States and maritime States, during the negotiation of the text of the LOS Convention, was related to the issue of sovereign rights and navigational rights through straits used for international navigation (George 2004). The LOS Convention seeks a balance between the two interests, through the regime of non-suspendable innocent passage in the territorial sea in non-strait waters and transit passage through the territorial sea in international straits.

The chapter by Manoj Gupta in this volume has dealt at length with the issue of how, from a geopolitical angle, effective forms of ocean governance to address SLOC security have been rendered problematic by the increased number of nation-states and their differing interpretations of the UN law of the sea. More recently, several significant international incidents in Asian exclusive economic zones have exposed the ambiguity and lack of agreement regarding the regime governing foreign military activities in coastal States' EEZs (see Valencia and Akimoto 2006). What follows next is not intended to be an account of the growing militarization of the Indian Ocean-space, but rather a comment on how increasingly influential the social construction of this ocean-space as a great void or force-field (rigorously pursued by the leading maritime/naval powers of today) is coming in conflict with the 'enclosures' outlined by the 1982 LOS Convention; especially the Exclusive Economic Zone (EEZ).

The Indian Ocean as a Force-Field: Maritime Zones, Passages and "Choke Points"

The Indian Ocean after nearly a decade-long strategic low-profile and invisibility is once again becoming an area of geo-strategic rivalry among a number of key powers and littoral states. The social construction of the Indian Ocean-space as a force-field or battlefield is being currently reinforced by what Don Berlin (2002, 2004) has described as the 'great base race' in the region. According to Berlin (Ibid.), the great base race carries enormous implications for Asia and beyond. He draws attention towards the emerging geographies of new strategic bases emphasizing forward defence. One important consequence of such factors at work, according to him, is the deepening of the identity and coherence of the Indian Ocean region as a geo-strategic space. According to Berlin (Ibid.):

> In terms of military space, these facilities are linking states (e.g. India and Malaysia) that in recent history were relatively separate from one another. At the same time, the advent and strategic reach of these installations will blur the boundaries and weaken the salience of some of Asia's traditional sub-regions; that is Northeast Asia, Southeast Asia, and South Asia. In so doing, and obviously in conjunction with forces not addressed here, the Indian Ocean region itself will grow, absorbing once peripheral zones. *As the ocean figuratively overflows its banks, capital cities from Asmara (Eritrea) to Dushanbe (Tajikistan) to Kuala Lumpur will be incorporated in various degrees into this enlarged global subsystem. This process could be driving the region towards a status reminiscent of the large, highly interactive zone that existed in the centuries before the final triumph of the West in these waters beginning in the mid-eighteenth century.* (emphasis added)

What Paul Bracken (1999: 168-170) has to say on 'new powers and the old order' is equally pertinent to the rejuvenated social construction of the Indian Ocean space as a force-field. According to Bracken the central challenge of the 21st century relates to the emergence of new technologically armed states, and their peaceful assimilation into the world community. This challenge, until recently, was defined and debated largely in geo-economic terms. Against the backdrop of Asian economies staging a comeback, the attention has increasingly been focused on whether and how China, India and others could be absorbed into the worldwide trading and monetary systems. Although the economic challenge remains, argues Bracken, it should now be seen in conjunction with the military one:

> The countries that must be brought into global economy have substantially greater military capacities. It is therefore no longer possible to dictate high-handedly what the entry conditions are. The problem here is not one that is subject to technical fixes; rather, it entails a newly constructed international system in which the West's military superiority no longer goes unquestioned and its capacity to backstop its vision of

world order is sharply limited compared to what it was even a few years back. Economic and military potentials have always been interlinked, and Asia's acknowledged economic rise is now leading to a military rise as well. ((Bracken, 1999: 169).

Looking ahead, what kinds of trends are probable in the international geopolitical economy, which might, directly or indirectly, influence the 'security' of Indian Ocean Sea Lanes of Communication? According to Wallerstein (2006: 94), over the next two decades or so "lurking behind any possible reconfiguration of world politics would be questions of access to energy and to water, in a world beset by ecological dilemmas and potentially producing vastly more than existing capacities of capitalist accumulation. Here could be the most explosive issues of all, for which no geopolitical maneuvering or reshuffling offers any solution." Furthermore, points out Wallerstein (2003), there are strong gales blowing from all directions against US hegemony. "Whether the ultimate outcome will be a less or more egalitarian and democratic order is totally uncertain. But the world that emerges will be a consequence of how we act, collectively and concretely, in the decades to come."

The challenge of acting 'collectively' and 'concretely' in the domain of SLOC security is likely to be compounded by a number of factors and forces that this collection of essays has addressed at some length. One among several critical issues, which are likely to acquire greater geopolitical as well as legal complexity in the times to come, is the one concerning navigation and overflight in the exclusive economic zone. It is useful to recall that one of the most controversial issues during the negotiation of the text of the 1982 UNCLOS was that of military activities in the EEZ. It continues to remain so in state practices. For example, some coastal states such as Bangladesh, Brazil, Cape Verde, Malaysia, Pakistan and Uruguay maintain that other states cannot carry out military exercises or maneuvers in or over their EEZ without their consent. They argue that such uninvited military activities could pose a serious threat to national security and undermine their resource sovereignty (see Valencia and Akimoto 2006). Whereas others take exactly the opposite stand. For example, the United States and other maritime powers insist, with all emphasis at their command, on ensuring the freedom of military activities in the EEZ on the grounds that their naval and air access and mobility could be severely compromised by what they perceive as a global EEZ enclosure movement.

The above mentioned disagreement has already caused several incidents in the EEZ's of the Asia-Pacific region, including the March 2001 confrontation between the US Navy survey vessel Bowditch and a Chinese frigate in China's EEZ. Looking ahead one can expect more such incidents regarding military and intelligence gathering activities in the EEZs of the Indian Ocean. A rather visible trend of coastal States attaching increasing importance to their control of the EEZ assumes greater impetus in the light of the fact that, "of the 1700 warships expected to be built during the next few years, a majority will be smaller, coastal patrol vessels and corvettes" (Valencia and Akimoto 2006: 705).

> In particular, intelligence gathering activities in the EEZs are likely to become more controversial and more dangerous. In Asia, this disturbing prospect reflects the increasing and changing demands for technical intelligence; the robust weapons acquisition programs, especially increasing electronic weapons acquisition programs, especially increasing electronic warfare capabilities; and the widespread development of information warfare capabilities. Further, the scale and scope of maritime and airborne intelligence collection activities are likely to expand rapidly over the next decade, involving levels and sorts of activities quite unprecedented in peacetime. They will not only become more intensive; they will generally be more intrusive. They will generate tensions and more frequent crises; they will produce defensive reactions and escalatory dynamics; and they will lead to less stability in the most affected regions, especially Asia. (Ibid.)

Correspondingly, it is to state the obvious perhaps that the EEZ is likely to provide a large number of conflicts and disputes over navigational rights and freedoms in the new century. This is the 'enclosure' of costal maritime jurisdiction where conflicting post-modern social constructions of Indian Ocean-space are going to be most pronounced. As far as the relevant UNCLOS III provisions are concerned, they leave much to be desired in respect of offering solutions to newly emerging problems. Article 58(3) provides that in exercising their rights and performing their duties in the EEZ:

> States shall have due regard to the rights and duties of the Coastal State and shall comply with the laws and regulations adopted by the coastal State.

Whereas Article 59 of 1982 UNCLOS, aiming at balancing rights and interests of States, reads as follows:

> In cases where this Convention does not attribute rights or jurisdiction to the coastal State or to other States within the exclusive economic zone, and a conflict arises between the interest of the coastal State or States, the

conflict should be resolved on the basis of equity and in the light of all the relevant circumstances, taking into account the respective importance of the interests involved to the parties as well as to the international community as a whole.

Even a casual reading of these provisions reveals that considerable ambiguity prevails over the meaning of 'due regard', other than 'relevant circumstances', and the respective importance of interests involved to the parties as well as the international community as a whole (Valencia and Akimoto 2006: 705). That the activity should not interfere with the 'rights' and 'interests' of the states concerned appears to be the one and only 'specific' criteria. But such a clause is of little practical utility and use in the absence of an agreement on what constitutes such rights and interests. Nor is there any agreement as to "whether the interference must be unreasonable or not, and whether it could be or must be actual or potential" (Ibid.).

The 'War on Terrorism' and the Security of IO Sea Lanes

Despite increasing optimism in sea-based commerce, as pointed out by Vijay Sakhuja (2004:270-71), sea-lanes continue to be vulnerable to disruption. These routes are increasingly threatened by piracy, drug trafficking, gun running, human smuggling and pollution. Moreover, it is equally important to note that, "none of these activities are independent of the other and failure in one often leads to failure in others" (Ibid.). Needless to emphasize perhaps that the ideal as well as imperative of international cooperation in this regard calls for a holistic approach and an acknowledgement by all concerned that, as mentioned in the Preamble to the 1982 UNCLOS, "the problems of the ocean space are closely interrelated and need to be considered as a whole".

The spectre of maritime terrorism, especially with regard to the Indian Ocean, has acquired additional visibility and urgency after 9/11. The manner in which the interplay between national security concerns and the marine geopolitical-legal environment is going to unfold in this regard is still not clear. Yet the implications that international responses to terrorism carry for the law of the sea need to be explored and assessed. Especially in the case of the Indian Ocean, it is neither possible nor desirable to ignore the changes and challenges confronted by the littoral States (especially the Islamic States) in this regard since the September 2001 terrorist attacks on the United States. What kind of state practices could possibly follow in response to terrorism? The following issues raised by Donald Rothwell (2004: 354) are likely to become increasingly important and compelling in the Indian Ocean region.

...clearly there is considerable scope for coastal States in the name of enhanced national security to take additional measures within their existing maritime zones to control the activities of terrorists and international criminals more generally. Such activities can range from increased surveillance and interdiction, closure of sensitive straits and navigation routes, assertions of extended maritime claims to the territorial sea or contiguous zones so as to enhance prescriptive jurisdiction, and navigational controls of all vessels passing through coastal State waters including the EEZ.

The challenge for the law of the sea and those States and institutions responsible for its maintenance will be to try and ensure the balance found in the LOS Convention is maintained. Greater vigilance will be required from bodies such as the IMO especially as regards the impact upon legitimate commercial shipping. However, while the challenges may be new in some respects the issues are not, as, especially in the case of navigational freedoms, there was much experience throughout the twentieth century in dealing with excessive regulation arising out of fears for coastal State security. What may be required however, is more attempts at maritime confidence-building especially at regional and sub-regional levels.

However, it is not only the State actors that are faced with the challenge of adopting additional measures to deal with terrorism. Non-state actors and market players too have started responding to the challenge in accordance with their own perceptions and interests. Following is an example of how these actors too are actively engaged in a new social construction of the Eurasia ocean-space.

London-based insurer Lloyd's dropped the Malacca Straits from its list of 20 war-risk areas for shipping prone to war, strikes, terrorism and related perils, on 7th of August 2006 a year after declaring the Straits a dangerous area. The Lloyd's Market Association Joint War Committee (JWC) expressed its satisfaction over what it described as a "significant improvement" in securing the Straits. However, it was resolved by the JWC that ships visiting northeast ports on the Indonesian island of Sumatra will continue to pay additional war-risk insurance premiums (although transit will be excluded). It is important to note that there are differences in perspective between the JWC and the three littoral states regarding security in the Malacca Straits. Whereas the JWC would like to calculate 'security' in the straits from the 'business angle', the littoral states might like to approach the same from the standpoint of sovereignty/authority, legitimacy and effectiveness. Malaysia's position, for example, has been that the statistics reflect the fact that area was safe from global terrorism.

Even though it usually takes a catastrophe of enormous proportions, such as a war or natural disaster, to impose premium hikes in the cargo market's soft cycle, competing social constructions of maritime risks will continue to invite arguments and counter-arguments from both the state actors and market players. In the wake of the Malacca Straits debate over the threat of terrorism attacks, a question has been raised (in certain quarters) whether the Strait of Hormuz would be the next major shipping lane of potential concern. During the Iran-Iraq War, the cargo insurance market did impose very high rates on certain high risk areas affected by the conflict.

In the pursuit of securing the sea lanes of communication the entry of private security companies has also raised several complex legal and political issues that would increasingly demand greater attention and analysis. PSCs based in Singapore have been providing niche armed guards and escort services for ships using the Malacca Strait. The operations of PSCs in the Malacca Strait do raise difficult policy issues, especially for Malaysia and Indonesia. Moreover, the stipulations of PSCs permits, littoral state permission for PSCs to operate in the Strait, actual PSC operations, and littoral jurisdiction and enforcement authority are highly ambiguous and controversial.

Indian Ocean-Space and SLOCs in the Era of Postmodern Capitalism: Beyond "Freedom" vs. "Enclosure"

Just as postmodern capitalism has intensified the Indian Ocean-style great void idealization, which was once the hallmark of the industrial-capitalist-era ocean-space construction, it has also intensified that era's tendency towards territorialization. Whereas during the industrial era, the territorialization of the sea was more or less confined to discrete coastal zones, more recently, there have been obvious trends towards territorialization of ever larger areas of ocean-space. The prospects of a conflict between the tendency to idealize the ocean as an empty space of flows, and a countervailing trend towards territorialization, are therefore steadily on the rise in the Indian Ocean region.

It is in within such a context (with humanity having entered a new era in its dealing with the sea) that we need to not only rethink and re-imagine 'sea lanes of communication' but also situate them in a broader context. The end of the East-West ideological Cold War geopolitics has radically altered the previous balance of sea power and the corporate globalization of international geopolitical economy is breaking down national borders in the shipping industry. Legal structures and norms established under the UNCLOS III too are being subjected to new social constructions of ocean-space. What one is witnessing in the maritime world is nothing short of a paradigm shift.

What Sam Bateman (2000: 109) has to say for the Asia-Pacific in general is equally valid and compelling in the case of the Indian Ocean region. According to Bateman, "the straits regime stands to become a more vital and possibly controversial issue in the geopolitics of the Asia Pacific. Conflicting interests are involved with, on the one hand, the environmental and sovereignty concerns of coastal States, and on the other, the essential movement of shipping. The issue needs to be addressed in the region at several levels, including by actions aimed at improving ship safety to prevent accidents occurring, such as improving standards of Port State Control and bridge-watch keeping, rather than by simply restricting rights and freedoms of navigation".

Concluding Observations

With mobility assuming an increasingly significant role in international geopolitical economy, and with new technologies becoming available for exploiting previously inaccessible resources, it is highly doubtful whether each dominant social construction of the Indian Ocean-space, on its own, will be able to provide effective and authoritative solutions to the problem of 'securing' the Indian Ocean sea lanes of communication. Besides being partial, individual social constructions, on their own, are unable to address cross-cutting issues in the domain of ocean governance; issues that straddle various institutional landscapes as well as policy domains. Various State as well as non-state actors, with a stake in ensuring the safety and security of Indian Ocean sea lanes of communication, will continue to work out their respective sea-lane defence mission. In all probability, various institutions, agencies and ministries within the countries concerned too will continue to conceive and construct the Indian Ocean-space rather differently; a point highlighted by Swaran Singh's contribution to this collection.

The challenge of a comprehensive security for the sea lanes/lines of communication needs to be approached and analyzed in the context of a growing convergence (bordering merger) between the "spaces of places" and "spaces of flows". At the same time, Indian Ocean-space needs to be re-conceptualized increasingly in amphibious terms. Land and sea routes, it may be argued, are often reciprocal, although they can also compete, or act as alternatives. In other words, there is a compelling need to look beyond the conventional, fairly well entrenched, conceptual binaries of oceanic-terrestrial and take a holistic and integrated view of the Indian Ocean-space. Much more difficult perhaps is to resolve what Philip E. Steinberg (1999: 254-55) aptly describes as the "contradictory tendency' in the dominant spatiality of modern-era marine governance: "both the tendency to enclose ocean space with lines of division and the tendency to construct it as a friction-free surface characterized by lines of connection".

It was pointed out in the introductory chapter to this volume that in order to secure the sea lanes of communication, a much more broadly-based maritime security strategy for the Indian Ocean region is needed, which, in turn, would necessarily demand consideration of a wide range of economic, environmental, political and social factors. It would also require greater inter-organisational collaboration within states and the treatment of "bureaucratic sclerosis" for its successful implementation. We would like to push this argument further by saying that in order to deepen such a collaboration, what is needed is a dialogue between diverse social-constructions of Indian Ocean-space as well as within each social construction itself. Inability or lack of will to do so can seriously jeopardize the prospects of both evolving and sustaining a consensus-based holistic understanding of how to secure the sea lanes of communication in and around the Indian Ocean.

References

Akimoto, K. (2006a), "How the Japanese and Global Economy would be Impacted by Potential Disruption of Oil Tanker Transport in Malacca Strait". Paper presented at the *Workshop on International Cooperation in the War Against Terror in the Asia-Pacific Region with Special Emphasis on the Malacca Strait,* Mississippi State University, 8-9 March.

Barendse, R.J. (2002), *The Arabian Sea: The Indian Ocean World of the Seventeenth Century*, New Delhi: Vision Books.

Bateman, S., Rothwell, D. R. and VanderZwaag,D. (2000), "Navigational Rights and Freedoms in the New Millennium: Dealing with 20th Century Controversies and 21st Century Challenges", in Donald R. Rothwell and Sam Bateman (eds.) *Navigational Rights and Freedoms and the New Law of the Sea,* The Hague: Martinus Nijhoff Publishers: 314-335.

Bateman, S. (2000), "The Regime of Straits Transit Passage in the Asia Pacific: Political and Strategic Issues", in Rothwell, D.R. and Bateman, S. (eds.) *Navigational Rights and Freedoms and the New Law of the Sea*, The Hague: Martinus Nijhoff Publishers: 94-109.

Berlin, D. (2002), "Indian Ocean Redux Arms, Bases and Reemergence of Strategic Rivalries", *Journal of Indian Ocean Studies,* 10(1): 26.

Berlin, D. (2004), "The 'Great Base Race' in the Indian Ocean Littoral: Conflict Prevention or Stimulation?" *Contemporary South Asia 13(3), (September): 239-255.*

Borgese, E. M. (1998), *The Oceanic Circle: Governing the Seas as a Global Resource, Tokyo*: United Nations University Press.

Bracken, P. (2000), *Fire in the East: The Rise of Asian Military Power and the Second Nuclear Age*, (First Perennial Edition), New York: HarperCollins Publishers.

George, M. (2004), "The Regulation of Maritime Traffic in Straits for International Navigation" in Oude Elferink, A.G. and Rothwell, D. R. (eds) *Ocean Management in the 21st Century: International Framework and Responses*: 19-48.

McPherson, K. (2004), *The Indian Ocean: A History of People and the Sea in Maritime India*, New Delhi: Oxford University Press.

Rothwell, D. (2004), "Oceans Management and the Law of the Sea in the Twenty-First Century" in Oude Elferink, A.G. and Rothwell, D. R. *(eds) Ocean Management in the 21st Century: International Framework and Responses*: 329-356.

Sakhuja, V. (2004), "Contemporary Piracy, Terrorism and Disorder at Sea: Challenges for Sea Lane Security in the Indian Ocean" in Dennis Rumley and Sanjay Chaturvedi (eds.) *Geopolitical Orientations, Regionalism and Security in the Indian Ocean*, New Delhi: South Asian Publishers: 269-281.

Smith, R. W. and Roach, J. A. (1997), "Navigation Rights and Responsibilities in International Straits" in Hamzah Ahmad (ed.) *The Straits of Malacca: International Cooperation in Trade Funding & Navigational Safety*, Jalan: Pelanduk Publications:

Steinberg, P.E. (2001), *The Social Construction of the Ocean,* Cambridge: Cambridge University Press.

Steinberg, P. E. (1999), "Lines of Division, Lines of Connection: Stewardship in the World Ocean", *The Geographical Review,* 89(2): 254-264.

Valencia, M. J. and Akimoto, K. (2006), "Guidelines for Navigation and Overflight in the Exclusive Economic Zone", *Marine Policy* 30: 704-711.

Wallerstein, I. (2006), "The Curve of American Power", *New Left Review*, 49(July-August): 77-94.

Wallerstein, I. (2003), "Emerging Global Anarchy", *New Left Review*, 22 (July-August): 27-35.

The Security of Sea Lanes of Communication in the Indian Ocean Region

INDEX WORDS